T0001687

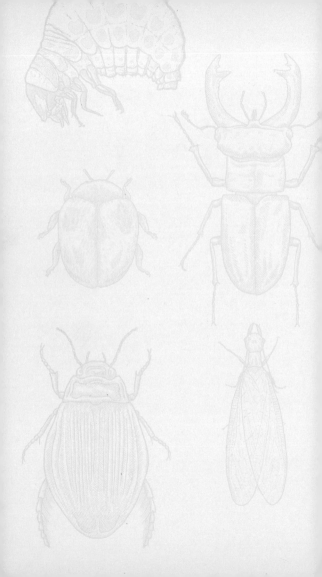

THE LITTLE BOOK OF
BEETLES

THE LITTLE BOOK OF
BEETLES

With color illustrations by Tugce Okay

ARTHUR V. EVANS

PRINCETON UNIVERSITY PRESS
PRINCETON AND OXFORD

Published in 2024 by Princeton University Press
41 William Street, Princeton, New Jersey 08540
99 Banbury Road, Oxford OX2 6JX
press.princeton.edu

Copyright © 2024 by UniPress Books Limited
www.unipressbooks.com

Library of Congress Control Number 2023943773
ISBN 978-0-691-25177-6
Ebook ISBN 978-0-691-25178-3

Typeset in Calluna and Futura PT

Printed and bound in China
1 3 5 7 9 10 8 6 4 2

British Library Cataloging-in-Publication Data is available

This book was conceived, designed, and produced by UniPress Books Limited

Publisher: Nigel Browning
Managing editor: Slav Todorov
Project development and management: Ruth Patrick
Design and art direction: Lindsey Johns
Copy editor: Caroline West
Proofreader: Robin Pridy
Color illustrations: Tugce Okay
Line illustrations: Ian Durneen

IMAGE CREDITS

Age Fotostock: 148 James E. Lloyd. **Dreamstime.com**: 29 Oleksii Kriachko;
48 Paul Reeves; 53 Alexmax; 100 Bennymarty; 125 Photogolfer.
iStock: 121 Henrik_L. **Nature Picture Library**: 11 Mark Moffett; 15 Thomas
Marent; 66 John Abbott; 75 Paul Bertner; 79 Nature Production; 88 Konrad
Wothe; 111 Michael Durham; 147 PREMAPHOTOS. **Shutterstock**: 139r alslutsky.
Other: 25 Anders L. Damgaard www.amber-inclusions.dk; 40 David R.
Maddison; 43 Joyce Gross; 63 Soebe; 83, 134 Arthur V. Evans;
105 UBC Micrometeorology from Vancouver, Canada; 115 John Abbott;
139l Yves Bousquet. **Additional illustration references**: 35 Mario Sarto/
Björn S...; 47 Alpsdake; 69 *Journal of Anatomy* 2018 Jun, 232(6): 997–1015,
figure 8, © 2018 Anatomical Society; 87 Bernard DUPONT; 95 P.S. Meng,
K. Hoover, and M.A. Keena/Patrick Randall; 99 Yves Bousquet;
107 Geoff Gallice; 131 Judy Gallagher.

CONTENTS

INTRODUCTION

Since high school, I have had an inordinate fondness for beetles. Growing up along the southwestern fringes of the Mojave Desert in California afforded me ample opportunities to explore diverse habitats rich in insect life. Family camping trips throughout the Golden State further exposed me to myriad species living in coastal and montane habitats, too.

IN THE BEGINNING

The intellectual framework for my passion began after a behind-the-scenes tour of the insect collection at the Natural History Museum of Los Angeles County. There I discovered cabinets filled with systematically arranged beetle specimens, each affixed with a label that provided important details of its capture. Beginning in high school and equipped with a freshly minted driver's license, specimen label data, and records gleaned from the scientific literature, I set out on field trips throughout southern California and Arizona in search of beetles. My goal was to find species new to me and, whenever possible, discover and describe species new to science. Eventually my research led to field trips and museum visits across the United States and Canada, the United Kingdom, parts of Mexico and Central America, and throughout southern Africa.

After completing my doctoral degree in systematic entomology at the University of Pretoria in South Africa, my interests expanded to include informal science education. For ten years I worked as the Insect Zoo Director at the Natural History Museum of Los Angeles

County, where I set out to encourage a greater awareness and appreciation of insects and other arthropods. I developed a traveling insect zoo that went out to schools, libraries, and summer camps, as well as a series of workshops for teachers that encouraged the inclusion of insects in school curricula.

SPREADING THE WORD

I have always welcomed opportunities to share my passion for all things insect, especially beetles. *The Little Book of Beetles* is a distillation of my experiences as both a scientist and educator. Its profusely illustrated chapters cover beetle evolution, diversity, structures, development, and habits. Another chapter explores the human-beetle interface through essays highlighting the roles of beetles in folklore and popular culture, as well as the inspiration they provide in the fields of science, technology, and medicine. The penultimate chapter focuses on the enjoyment and study of beetles and the need to conserve them, while the final chapter offers up fun facts spotlighting some of the amazing and quirky aspects of Coleoptera.

The ubiquity of beetles, combined with their diversity of form and behavior, have nourished and sustained my intellectual curiosity for more than half a century. Over the course of my career, I have learned that there are two types of people—those who are fascinated by beetles and those who don't yet know they are fascinated by these amazing animals. For the initiated, *The Little Book of Beetles* will provide a fresh look at beetles from around the world, while those just beginning their exploration of these insects will find this book a colorful and engaging introduction. It is my hope that *The Little Book of Beetles* will help inspire a new generation of naturalists and scientists.

Arthur V. Evans, D.Sc.

EPITOME OF DIVERSITY

With more than 400,000 species, a number that is more than ten-fold the total number of all vertebrates (fish, amphibians, reptiles, mammals, and birds), beetles are the largest group of animals on Earth. On land, they are unmatched in their range of size, form, and color. The astonishing diversity of beetles is attributable to their ancient body plan. For more than 285 million years, their distinctively compact and armored bodies have evolved and adapted to the countless challenges presented by ever-changing environments. The behavioral, developmental, and physiological attributes of beetles have also enabled them as a group to survive and thrive in innumerable terrestrial niches, whether high atop the canopies of steaming rain forests or in blistering and seemingly desolate deserts. Several groups are also adapted to living on or in bodies of fresh water.

INSPIRATION

Familiar yet alien, the intriguing and sometimes bizarre behaviors of beetles have long fascinated humans, thus influencing cultures and inspiring myths and legends for thousands of years. In 19th-century Europe, interest in beetles was firmly established among commoners and aristocrats alike. They scoured the countryside and visited foreign lands to collect specimens and systematically build enormous collections of pinned and pickled beetles in an attempt to understand the very underpinnings of God and nature.

INNOVATION

Today, scientists worldwide working in multiple fields are conducting meticulous studies of beetle structures, behaviors, and defensive secretions using innovative technologies, including molecular analyses that promise a wealth of important scientific, technological, and medical advances for generations to come.

ACROCINUS LONGIMANUS

The harlequin beetle, *Acrocinus longimanus*, inhabits tropical forests from southern Mexico to Uruguay. The striking color pattern of their elytra is reflected in the common name of these sap-feeding beetles. Males are distinguished by their extremely long forelegs that they use to grip tree trunks as they butt heads with rival males. Pseudoscorpions in search of new feeding and mating sites regularly hitch rides on these large beetles, a form of *symbiosis* known as phoresy. The larval tissues of these beetles contain antifungal peptides that may provide a treatment for life-threatening infections acquired by patients in hospital settings.

↑ Although harlequin beetle larvae normally develop in fig tree trunks, they are serious pests of breadfruit trees introduced from Southeast Asia.

PERFECTLY APPOINTED

T he bodies of beetles, along with those of other insects and arthropods (crustaceans, arachnids, millipedes, centipedes, and their kin), are bound within a tough, segmented *exoskeleton*. These segments, along with their jointed antennae, mouthparts, and legs, are joined together by more or less flexible hinges that afford them great mobility. As in all insects, the segments of the beetle exoskeleton are grouped into three major body regions: head, thorax, and abdomen.

Functioning as both skin and skeleton, the beetle exoskeleton is appointed with a wide array of structures that enable them to chew, walk, run, burrow, swim, and fly with ease. It also serves as a platform for sensory structures that help beetles to perceive the sights, smells, and sounds around them, as well as provide them with the sense of touch. The beetle exoskeleton also comes in an astounding array of colors and patterns that help them to find mates, avoid danger, and regulate their body temperature.

↓ Male European stag beetles (*Lucanus cervus*) have antler-like mandibles.

↓ Female great diving beetles (*Dytiscus marginalis*) have distinctly grooved *elytra*.

→ The royal goliath beetle, *Goliathus regius*, inhabits tropical forests in West Africa. Males are distinguished by the Y-shaped horns on their heads that are used in battle with other males over potential mates or feeding sites.

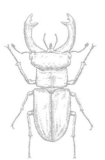

RECIPE FOR SUCCESS

The evolutionary success of beetles, as measured by their sheer number of species and ecological diversity, is unmatched by any other group of terrestrial animals. Their pre-Cretaceous origins, coupled with their physical and physiological attributes, contributed to the evolution and longevity of numerous modern lineages that arose and diversified during the Jurassic, 201–145 Mya (Mya = million years ago). Based on the fossil record, ancient beetles were mostly small and capable of flight; this allowed them to avoid predators and cover great distances while searching for food and mates. The evolution of their uniquely hardened forewings, or elytra, in combination with the *subelytral cavity* immediately underneath, preadapted beetles for living in all kinds of terrestrial and aquatic habitats, thus enabling them to exploit ecological niches unoccupied or underutilized by other organisms.

ELYTRA

The elytra help protect the abdomen and the organ systems contained within. They also shield the delicate hindwings (folded underneath) from abrasion while beetles burrow into the soil or wedge themselves under loose bark. The subelytral cavity has enabled terrestrial beetles to adapt to living in harsh, dry conditions by helping them to regulate their body temperatures and conserve moisture. In aquatic beetles, the subelytral cavity provides a place to store a bubble of air from which they can extract oxygen to breathe while underwater.

↑ The thickened forewings of beetles are called elytra.

DEVELOPMENT

Beetles evolved as part of a branch of insects that undergo *holometaboly*. They and other holometabolous insects (for example, ants, bees, wasps, flies, fleas, butterflies, and moths) pass through four distinct life stages: egg, larva, pupa, and adult. Beetle larvae are substantially different from the adults not only in form but also in terms of their food preferences and ecological requirements. Holometaboly in beetles and other insects not only reduces competition between parents and their offspring for food and space, but also better adapts species overall for surviving seasonal temperature fluctuations, especially in temperate climates.

BRANCHING OUT

The diversification of most modern lineages of beetles peaked during the Triassic (252–201 Mya) and Jurassic, well before the appearance of angiosperms (flowering plants). By the time angiosperms had appeared, plant-feeding beetles were already capable of producing important digestive enzymes that were likely acquired through horizontal gene transfer from bacteria and fungi. This evolutionary development shifted reliance on symbiotic fungi and bacteria in their gut for digestion to symbiotic-independent mechanisms. Over time, these herbivorous beetles became more efficient at digesting plant tissues, thus setting the stage for increasing degrees of food plant specificity and paving the way for leaf and stem mining, wood boring, specialized mycophagy, and other uniquely specialized feeding habits.

BORROWED GENES

It was previously thought that beetle diversity went into overdrive at the beetle–plant interface beginning in the Late Mesozoic, around 66 Mya, and that the ever-increasing diversity of angiosperms led directly to the evolutionary radiation of beetles. In order to digest plant tissues, it was believed that beetles were heavily reliant on endosymbiotic microorganisms. However, recent studies on how beetle genes interact with each other and their environments (genomics), coupled with investigations into their digestive physiologies, suggest that many lineages of herbivorous beetles were preadapted to take full advantage of the appearance of angiosperms and their novel vegetative structures.

BEETLES AS PESTS

Plant and animal materials improperly stored in pantries, warehouses, and museum collections are likely to attract and nourish beetles considered to be pests. Other pestiferous beetles attack garden plants and crops, or destroy forests managed for lumber and other timber products. Bark and ambrosia beetles that usually attack dead, injured, or felled trees can decimate ornamental and forest trees, especially those stressed by overwatering or drought. The tunneling activities of these and other beetles injure or kill living trees by disrupting the tree's ability to transport water and nutrients and introducing potentially lethal infections. Select species of deathwatch and powderpost beetles that attack only dead wood can damage wood carvings, furniture, flooring, and paneling.

Insecticides that kill specific pests, such as neonicotinoids, are used on crops to increase profits. However, some researchers suggest that applications of neonicotinoids not only reduce bee populations worldwide, but also harm other beneficial insects essential to agriculture, including beetles. For example, laboratory and field experiments demonstrate that slugs unaffected by neonicotinoids applied to soybeans harbor enough of the chemical in their tissues to harm or kill beneficial ground beetles that eat slugs and other pests. Declining beetle populations led to increased numbers of slugs that consume seedlings, thus reducing the numbers of plants and lowering total crop yield.

ECONOMICS

Herbivorous beetles provide valuable ecological services by breaking down and recycling nutrients bound up in living and dead plant tissues. However, when select species focus their attentions on wood products or plants growing in gardens, parks, forests, and viewsheds, the aesthetic and economic impacts can be severe. Worldwide, catastrophic monetary losses resulting from lost agricultural production, damaged goods, and killed forest trees costs hundreds of billions of dollars annually. The loss of forests due to beetle damage can have serious long-term consequences by disrupting or destroying invaluable ecological services upon which we depend for clean air and water, erosion control, and nutrient cycling. Such losses are compounded by the enormous cost and long-term ecological damage that results from efforts to control beetle pests with pesticides.

← Native to North America, the Colorado potato beetle, *Leptinotarsa* *decemlineata,* is one of the most invasive leaf beetles in the world.

USEFUL BEETLES

In spite of the significant economic damage caused by relatively few species, most beetles are either of little or no direct consequence to us, or are enormously valuable for the services they provide. Examples of beneficial beetles include species deployed in museums for cleaning skeletons and those used as biological control agents.

TO THE BONE

Museums utilize flesh-eating skin or museum beetles in the genus *Dermestes* to clean skeletons for study as well as for exhibits. Animals ranging from small birds to giant whales are placed in enclosures teeming with museum beetles and their larvae, which will gnaw at the carcasses until nothing remains but skeleton. This process may take only a few days for smaller animals or several months for large marine mammals.

HEROES

Biological control involves the use of a pest's natural enemies (predators, *parasitoids*, herbivores, and pathogens) as control methods, rather than solely relying on pesticides that may adversely affect other animal and plant life. Modern biocontrol began with efforts in California to combat the cottony cushion scale, *Icerya purchasi*, an Australian insect that threatened to destroy that state's nascent citrus industry. Several species of Australian lady beetles that prey upon these scales were sent to California in the late 19th century and released to combat the citrus pest. Soon, the vedalia beetle, *Novius cardinalis*, was credited with saving California's citrus industry, and its use to control the cottony cushion scale was hailed as a miracle of science.

← Vedalia beetles (*Novius cardinalis*) are densely *pubescent*, red and black ladybird beetles.

SANITATION ENGINEERS

Dung beetles are also used as biocontrol agents. Australia's native dung beetle fauna prefers the mostly small, fibrous pellets produced by native marsupials and not the big, wet flops produced by cattle that serve as breeding sites for pestiferous flies and *parasites*. The accumulation of dried cow chips (cowpats) also reduces the amount of palatable cattle forage. To combat the annual hordes of flies and loss of pasturage, the Australian government imported dung beetles from southern Africa in the 1960s to bury cattle waste. Strict quarantine measures were put into place to avoid the introduction of parasites and other cattle pests into Australia. Once established, the dung beetles buried the cow dung, thus eliminating fly breeding sites while recycling nutrients and restoring pasture quality. Although the Australian Dung Beetle Project ended more than 35 years ago, 40 introduced dung beetle species still remain hard at work there.

RECLAMATION

Using beetles as biocontrol agents is not limited to harnessing the appetites of insect predators or scavengers of carrion and animal waste. Herbivorous species are sometimes pressed into service to help control pestiferous plants in wildlands in an effort to restore native habitats. For example, the rapid expansion of Eurasian tamarisk or saltcedar (*Tamarix* species) in both natural and artificial *riparian* habitats in western North America over the past two centuries has been directly linked with the decline of cottonwood–willow woodlands (*Populus* and *Salix*), mesquite bosque (predominantly *Prosopis*), and other native plant complexes. Groves of tamarisk not only displace native plant communities and wildlife, but they also increase soil salinity by diminishing local water sources. In an effort to reclaim tamarisk-dominated habitats for native riparian woodlands, entomologists introduced several herbivorous beetles, including the splendid tamarisk weevil, *Coniatus splendidulus*, and the northern tamarisk beetle, *Diorhabda carinulata*.

UNIQUE PERSPECTIVES

Just as more pixels of light improve the clarity of a digital image, so does studying the seemingly infinite variety of beetles enhance our understanding of the natural world. Research that is focused on beetles affords scientists a view of life that is unmatched by the investigation of most other groups of organisms. It is no coincidence that one of the major pillars of scientific thought, the theory of evolution via natural selection, was conceived independently by two Victorian naturalists, Charles Darwin (1809–1882) and Alfred Russel Wallace (1823–1913), who drew inspiration, at least in part, from their fascination with beetles. As our technology continues to advance, so will our ability to unravel the mysteries of beetle evolution. Locked away somewhere in the beetle genome are answers to questions that we have yet thought to ask; answers that will undoubtedly provide important insights into all life on Earth past, present, and future.

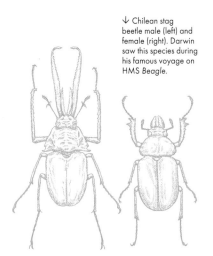

↓ Chilean stag beetle male (left) and female (right). Darwin saw this species during his famous voyage on HMS *Beagle*.

→ Alfred Russel Wallace found long-armed chafers (*Euchirus longimanus*) sapping on sugar palms in Indonesia, one of the locations in which he observed evidence supporting the theory of evolution via natural selection. Male chafers have incredibly long forelegs that are used in battle with other males.

ANCIENT ORIGINS

The ancestors of modern beetles, or Protocoleoptera, resembled modern Megaloptera (dobsonflies, fishflies, and alderflies) and some Neuroptera (antlions, lacewings, and owlflies), but had thickened forewings. Protocoleopterans appeared in the fossil record during the Carboniferous (around 327 Mya) and had disappeared by the Middle Triassic.

Coleopsis archaica, the earliest definitive beetle fossil, was discovered in Early Permian (around 285 Mya) deposits. This and other early coleopterans are distinguished from their protocoleopterans ancestors by their flatter, tougher, and more compact bodies, as well as by their shorter legs and antennae. Their elytra lacked any traces of venation and fitted snugly over the body.

All four extant suborders of beetles (Archostemata, Myxophaga, Adephaga, and Polyphaga) were present by the middle of the Triassic (around 235 Mya). The diversification of most modern lineages of beetles within these suborders peaked during the Triassic and Jurassic. At the close of the Jurassic (around 151 Mya), all major lineages known today above the family level were present.

↓ The extinct family †Tshekardocoleidae includes some of the oldest known beetle fossils that date back to the Permian.

Fossils that feature structures other than just elytra are †*Sylvacoleus sharovi* (left) and †*Moravocoleus permianus* (right).

→ The primitive family Cupedidae, commonly known as reticulated beetles, is represented by 8 genera and 37 species found worldwide, 4 of which occur in North America. *Tenomerga cinerea* is found in the hardwood forests of Ontario, Canada, and the eastern United States.

FOSSILIZATION

Two-dimensional beetle fossils consist mostly of compressed elytra and other fragments, or their impressions, found within layers of sedimentary rock. With most of their molecular structures and original colors lost, compression fossils preserve the fine structures of ancient beetles via carbon residues. Impression fossils offer casts of the beetle's form, with the level of detail depending on the physical quality of the surrounding sediment. Subfossils (remains not replaced by minerals) preserved in recent (23 Mya to about 11,000 years ago) deposits of river sediments, peat, and asphalt often include beetle species still extant.

FOSSILS IN 3D

Ancient beetles are also found inside petrified globs of tree sap called amber or copal. Three-dimensional inclusions of beetles and other organisms are preserved in incredible detail, often with portions of their original colors and tissues, including bits of subcellular details like DNA, partially intact. The bodies of petrified beetles found inside concretions are sometimes partly or wholly replaced with minerals and are preserved with great fidelity.

Using sophisticated scanning equipment, scientists discovered remarkably preserved beetle fossils inside 230-million-year-old coprolites (petrified dinosaur dung). Examination of petrified dinosaur feces and their three-dimensional beetle inclusions sheds light on both insectivorous dinosaur-eating habits and beetle evolution that predates amber.

ANCIENT VESTIGES

Ichnofossils, traces of ancient beetle-feeding activities on fossilized leaves, wood, and other plant tissues, reveal the most about the lives of ancient beetles. The distinctive subterranean nests of dung beetles—complete with brood balls—have been found preserved in association with herbivorous dinosaur nests and their coprolites.

↓ An *Apion* weevil (Brentidae) preserved in 50 million year-old Baltic amber. Beetles and the remains of other organisms preserved in amber are called inclusions.

THE FOSSIL RECORD

Insects are among the most ubiquitous and abundant groups of animals on Earth and their fossil record dates back to the Early Devonian (around 400 Mya). Due to the toughness of their exoskeletons, especially the hardened elytra, beetles are among the best represented among the insect fossil record that stretches across seven continents. All four extant suborders of Coleoptera are represented in the fossil record that includes preserved examples of nearly 70 percent of all known families, both extinct and extant.

NORTHERN BIAS

Most known beetle fossil deposits occur in the Northern Hemisphere, a fact attributable to the intensive exploration and study in Europe and North America, where the roots of paleoentomological investigations first took hold. Some of the richest and best-known beetle fossil deposits occur in China, Germany, Russia, and the United States. Arid regions around the world that have experienced geological uplifting often have fossil beds exposed by erosion and/or sparse vegetation. The relative lack of fossils from the Southern Hemisphere is likely due, in part, to the deposits there being covered in dense tropical vegetation.

AMBER VS. LACUSTRINE

Most beetle fossils are preserved as resinous inclusions or sealed in sedimentary deposits associated with various bodies of water. Amber and lacustrine deposits are especially rich in terms of their abundance and taxonomic diversity of beetle fossils. Lacustrine deposits dating from the Permian (300–252 Mya) are richest in fossil beetles. They are associated mostly with ancient lakes and marine margins and preserve the most specimens. Fossils preserved in older lacustrine deposits dating back to the Late Permian are the most instructive for elucidating macroevolutionary trends in Coleoptera.

The relatively small number of beetles preserved in amber only date back to the Early Cretaceous (145–100 Mya). Although preservation is vastly superior to that of other types of fossilization, beetle inclusions in amber are relatively recent and don't reveal the earliest stages of beetle evolution. Nearly all of the amber originating from the Mesozoic (252–66 Mya) and most of the Cenozoic (66 Mya to present) is from conifers. The largest deposit of amber in the world is found along the southern coast of the Baltic Sea and likely came from pines. Deposits of the equally well-known but more recent Dominican amber were formed by leguminous tree sap.

AGE

The geological age of fossilized beetles and other organisms, regardless of their preservation, is inferred by surrounding layers of rock, or strata, and other associated fossils of known age. Geologists analyze radioactive decay that naturally occurs in select elements such as potassium to determine the ages of Mesozoic and Cenozoic strata and other geological features.

↓ This Eocene (56–35 Mya) fossil of *Pulchritudo attenboroughi* (Chrysomelidae) has an unusual elytral pattern.

RADIOCARBON DATING

Determining the age of fossils less than 50,000 years old usually involves radiocarbon dating. Beetles and other organisms absorb carbon from their food and surrounding environment, including the radioactive carbon-14 isotope. Absorption ceases after death and the radioactive carbon present decays at a predictable rate. Measuring the remaining isotope gives an estimate as to how long the organism has been dead. The amount of carbon-14 available for absorption varies geographically and over the course of time. Thus, the results of carbon-14 dating must be calibrated with additional data sets extracted from various sources, including trees, lake and ocean sediments, stalagmites, and corals.

RECONSTRUCTING PHYLOGENY

Aristotle (384–322 BC) first recognized beetles as a group based on their possession of elytra. In 1758, Carl Linnaeus (1707–1778) incorporated Aristotle's concept of Coleoptera into his classification of animals. His and other early-19th-century classifications were organized on the basis of physical similarity. By the mid-19th century classifications of beetles and other organisms began to be arranged in a more linear fashion, with taxa thought to represent lower or more primitive forms followed by those taxa considered to be higher or more advanced.

NEW SYNTHESIS

Beetle classifications during the first half of the 20th century were constructed with the extensive use of a wide range of character systems. Wing venation and folding patterns, male genitalia, and larval morphology were incorporated into systematic studies of beetles in an attempt to construct genealogical classifications that truly reflected their evolutionary relationships. During this time, based on vast amounts of morphological data, the first three of the four suborders of Coleoptera (Archostemata, Adephaga, and Polyphaga) were recognized.

CLADISTICS

German fly specialist Willig Hennig (1913–1976) published *Grundzüge einer Theorie der phylogenetischen Systematik* in 1950. In this revolutionary work, he recognized that ancestral characters (plesiomorphies) do not infer relatedness, while shared advanced or derived characters (synapomorphies) do infer evolutionary relationships. His rigorous phylogenetic method is known today as cladistics. The reconstruction of phylogenetic branching patterns are generated by cladistic analyses of the physical, biochemical, behavioral, and zoogeographical attributes of modern beetles, as well as the structural peculiarities and distributions of fossils.

↑ Stag beetles (Lucanidae), such as the European *Lucanus cervus*, are related to scarabs. Males are often characterized by enormous mandibles, as shown here.

CLADOGRAMS

Cladistic analyses are used to reconstruct the evolutionary relationships of beetles and other organisms in a cladogram. The branching patterns of cladograms symbolize these hypothetical relationships and are used as guides for constructing classifications with taxa grouped by their shared novel characters or synapomorphies. Elytra are a major synapomorphic feature found in most adult beetles although, over the course of evolution, they have been lost in some species, usually in the females. Given their geological age and diversity, beetles embody an incredible wealth of morphological and molecular data that might one day reveal the causes and correlates of diversification.

MASTERS OF ADAPTATION

Past hypotheses accounting for the exceptional diversity of Coleoptera have focused on their unique morphological and developmental features, along with their co-evolution with angiosperms, as the primary drivers of their evolutionary success. Current investigations identify the geological age of beetles, coupled with their low extinction rates and key physiological attributes, as major contributing factors.

AGE AND LOW EXTINCTION RATES

The evolutionary success of beetles is partially due to their pre-Cretaceous origin and the durability of numerous modern lineages that arose and diversified during the Jurassic. Multiple shifts among several lineages of beetles from terrestrial to aquatic lifestyles are also a key factor. Analyses of older fossils are typically used to infer

---

FOOD PLANT SPECIFICITY

Genomic investigations exploring the interaction of beetle genes with each other and their environments, coupled with investigations into their digestive physiologies, suggest an alternative hypothesis to the idea of borrowed genes. Ancient plant-feeding beetles once reliant on bacteria and fungi in their guts for digestion began to evolve mechanisms independent of these symbiotic microorganisms. With the aid of critical digestive enzymes acquired via horizontal gene transfer from bacteria and fungi, herbivorous beetles became more efficient at digesting plant tissues, setting the stage for increasing degrees of food plant specificity. The appearance of angiosperms and their evolutionarily novel tissues contributed to the diversification of phytophagous beetles as evidenced by the evolution of their leaf and stem mining, wood boring, and other uniquely specialized plant-feeding behaviors.

---

evolutionary trends at the family level or above, while beetle fossils recovered from the Pliocene and Pleistocene (5 Mya to 11,000 years ago) offer better species-level data. Represented by mostly modern genera and species, these relatively recent deposits are a testament to the resilience and longevity of beetles species.

MORPHOLOGY AND DEVELOPMENT

Small and capable of flight, ancient beetles could cover greater distances to search for food and mates, escape predators, and exploit new habitats. The evolutionary innovation of snug-fitting elytra afforded better protection for their flight wings and abdomen. It also created a sealed chamber called the subelytral cavity that provided insulation for terrestrial beetles and a place to store oxygen in aquatic species.

The evolution of holometabolous development in beetles and other insects is characterized by adults and larvae with substantially different habits and ecological requirements. Holometaboly not only effectively reduces competition between adults and their offspring, it also better adapts species for living in climates with distinct seasons marked by fluctuating temperatures or rainfall.

REVISTING THE BEETLE–PLANT INTERFACE

The "beetle–plant interface" that began in the Late Mesozoic (around 66 Mya) has long been suggested as the genesis of one of the greatest drivers of beetle diversity. It was thought that the ever-increasing diversity of flowering plants led directly to the evolutionary radiation of phytophagous beetles as they evolved ever more feeding specializations with the aid of symbiotic microorganisms to assist them with the digestion of diverse plant tissues. However, there was little direct evidence to provide a consistent link between feeding on flowering plants and beetle diversification. Current evidence suggests that most modern lineages of beetles at the series and superfamily levels peaked during the Triassic and Jurassic, long before the appearance of angiosperms.

← Mitochondrial DNA extracted from beetle leg tissues from old museum specimens and subfossils can be sequenced for phylogenetic study.

PAST INFORMS
PRESENT AND FUTURE

Cladistic analyses of beetles incorporate molecular and morphological characters and other data sets, including the structural peculiarities and distributions of their fossilized remains. This research helps to reveal the evolutionary drivers that contributed to the extraordinary diversity of beetles. Additionally, changes in species distribution and composition preserved in the fossil record during times of climatic turmoil may provide indications as to how modern beetle populations will react to similar conditions in the Anthropocene, an unofficial unit of geologic time that began when humans first started having significant impacts on the environment.

The existing fossil record is woefully incomplete and there is much work to be done. Our knowledge of prehistoric beetles is largely based on fossils from the Northern Hemisphere. Newly discovered deposits elsewhere in the world will undoubtedly produce specimens that challenge our current assumptions about beetle phylogeny. Add to these the countless specimens of modern species already tucked away in the world's museums that await analysis and description. So many beetles, so little time.

← Strepsipterans (top) and neuropteroid insects (bottom) are considered the nearest living relatives of beetles and are frequently used as outgroups in cladistic analyses seeking to develop hypotheses regarding the evolutionary relationships of the major lineages of Coleoptera.

→ Several phylogenies have been proposed hypothesizing the evolutionary relationships of the suborders of Coleoptera. As taxon and gene sampling techniques improve, larger data sets will be analyzed that will refine current hypotheses on the geologic age and evolutionary relationships of beetles.

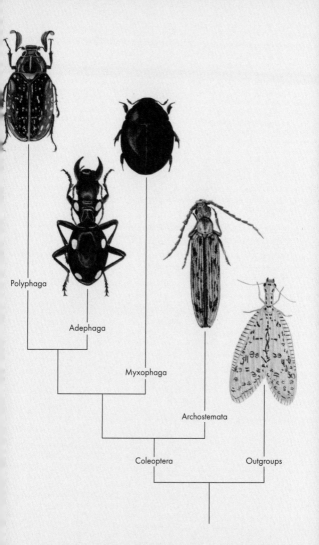

Polyphaga

Adephaga

Myxophaga

Archostemata

Coleoptera

Outgroups

33

ORDER OUT OF CHAOS

Since Aristotle, humans have attempted to understand biological diversity by naming organisms and classifying them based on their physical and other attributes. Arranging nature by naming and categorizing organisms became known as taxonomy, a word derived from the Greek words *taxis*, or "arrangement," and *nomos*, meaning "law." In short, taxonomy is our filing system for biological diversity. Taxonomy involves identifying, naming, describing, and classifying beetles and other organisms into a system of hierarchical ranks called taxa. Ideally, a beetle species is established on the basis of a group or population of interbreeding individuals that share a unique evolutionary history. Each species is distinguished from their nearest relatives by a unique combination of morphological, behavioral, distributional, ecological, and biochemical characteristics. Similar species are grouped into additional taxa (genera, tribes, families, etc.) that reflect their evolutionary relationships.

Taxonomy is part of a larger endeavor known as systematics that seeks to understand the diversity and interrelationships of organisms based on their shared evolutionary history, or phylogeny.

↓ Two beetle species mentioned by the Greek philosopher Aristotle in his *History of Animals* written in the 4th century BC were *kantharos* (*Scarabaeus sacer*), shown on the left, and *kleros* (*Trichodes apiarus*), on the right.

→ Aristotle also mentioned melolonthe (*Melolontha melolontha*) that is known today in English as the cockchafer. Adults of this European species emerge in spring. They prefer to lay their eggs in fields. Larvae feed on roots and will pupate in three or four years.

RESOLVING THE
BEETLE TREE OF LIFE

Carl Linnaeus established the system of naming beetles and other organisms using only two words, or a binominal—the genus and specific epithet. Written together and italicized with only the genus capitalized, the binominal serves as the beetle's formal scientific name, a unique identifier recognized the world over.

THE CODE

Names of Latin or Greek derivation are assigned following the rules laid out in the International Code of Zoological Nomenclature, or the Code. The Code includes criteria for the formation and viability of scientific names for beetles and other animals, the publication of those names and their descriptions in the scientific literature, and the subsequent usage of properly proposed scientific names in future publications. Scientific names are followed by the surname of the person who described the species. When a previously described species is moved to another genus, parentheses are placed around the author's name to indicate the new combination. The Code establishes that a holotype, the specimen upon which a new species description is based, be designated as the physical record for that species.

BEETLE HYPERDIVERSITY

With nearly 400,000 species constituting nearly 25 percent of animal diversity on Earth, beetles represent a hyperdiverse monophyletic clade in the tree of life. A robust beetle phylogeny is critical to our understanding of the processes underlying their evolutionary success as evidenced by their diversity. In spite of broad agreement on the existence of the four extant suborders of beetles examined in the following pages, there is still considerable debate regarding the interrelationships among them and the lineages within.

ESTABLISHING A HIERARCHY

Identifying and naming beetles are just the first steps toward building a classification that helps us to store and retrieve information, as well as to understand evolutionary relationships. Based on their shared characteristics, beetle species are first placed in genera (plural of genus), then genera into subtribes, subtribes into tribes, tribes into subfamilies, subfamilies into families, families into superfamilies, and superfamilies into suborders, all of which are classified in the order Coleoptera. Subtribes (-ina), tribes (-ini), subfamilies (-inae), families (-idae), and superfamilies (-oidea) are among the taxa that have universally accepted suffixes.

↑ Carl Linnaeus established modern taxonomy, see page 28.

HOMOLOGIES

Ideally, beetle classifications are not established on the basis of similarity but rather on shared derived characters known as synapomorphies. Synapomorphies are based on the examination of shared homologous characters. Homologous characters are identified by their similarity in appearance, as well as location and function. The sharing of synapomorphies implies a common genetic basis, and thus a shared evolutionary history among the species that possess those synapomorphies. Classifications based on synapomorphies have predictive value because they can give clues as to the qualities of lesser-known taxa based on their nearest relatives.

NATURAL CLASSIFICATIONS

The phylogenetic method by which characters of taxa are analyzed to develop hypotheses on the evolutionary relationships based on synapomorphies is called cladistics. Cladistic analyses include the examination of physical features, distribution, behavior, fossils, and embryology, as well as DNA. The results of cladistic analyses are expressed graphically as a branching diagram, or cladogram. Groups of branches, or clades, composed of hypothetical ancestors and their descendants are referred to as monophyletic. Classifications based on monophyletic clades best reflect our hypothesis of the natural or evolutionary relationships of the taxa examined.

ARCHOSTEMATA

The suborder Archostemata is a small group of Coleoptera that comprises 45 extant species in the families Cupedidae (widespread), Ommatidae (Neotropical, Australian), Micromalthidae (Nearctic, Neotropical), and Crowsoniellidae (Palearctic). Archostemata were most diverse during the Mesozoic when Cupedidae, Ommatidae, and Micromalthidae occurred in the Palearctic.

The biology and ecology of most archostematans remains largely unknown. Known larvae are long and slender with six short legs and associated with fungusy hardwoods and conifers. Adult cupedids and ommatids are sometimes attracted to lights. Nothing is known of Crowsoniellidae, other than only three adult specimens of the single species *Crowsoniella relicta* that were found washed from soil beneath a chestnut tree in Italy. The life cycle of micromalthids, represented by its sole species *Micromalthis debilis*, is utterly bizarre and among the most complex known of all beetles. Capable of two forms of asexual reproduction, the larvae appear to have a central role in the propagation of the species.

↓ The archostematan *Omma stanleyi* (Ommatidae) is a "living fossil" found under eucalyptus bark in eastern Australia.

↓ *Crowsoniella relicta* was named after one of the greatest coleopterists of the 20th century, Roy A. Crowson (1914–1999).

→ Adults of *Micromalthis debilis* (Micromalthidae) are rarely encountered. Widespread in the eastern United States, this species is also known from British Columbia, California, Belize, and other parts of the world, likely as a result of commerce.

MYXOPHAGA

Myxophaga was the last of the four beetle suborders proposed and is comprised of about 120 species classified in four small families, including Lepiceridae (Neotropical), Torrindincolidae (Afrotropical, Neotropical), Sphaeriusidae (all continents except Antarctica), and Hydroscaphidae (Afrotropical, Nearctic, Neotropical, Oriental, Palearctic). Despite its relatively small size, the suborder has an outsized role in elucidating the evolutionary history of Coleoptera.

Myxophagans are characterized by their small size, algae-feeding habits, and preference for mostly hygropetric habitats. The largest species measure $^1/_{10}$ in (2.7 mm) in length, but most species are much smaller. Hygropetric species typically inhabit the thin layers of water that cover rocks, algae, and aquatic plants found along streams and rivers, especially in warm, humid climates. Adults and larvae occupy interstitial spaces in wet sand and gravel along rivers, algal-covered seeps and stream banks, or spray zones associated with waterfalls. Adults are also occasionally found in association with flood debris or among tropical leaf litter some distance from water.

BEETLE FAMILY: SPHAERUSIDAE

The 22 species of Sphaerusidae are among the smallest and least studied of Myxophaga. These hygropetric species are distinguished mostly on the basis of their size and surface sculpturing, but recent molecular studies suggest that cryptic species with similar external features are likely. Specimens are seldom collected and little is known of their natural history or distribution. Based on comparisons of modern-day *Sphaerius* with two extinct genera (*Bezesporum* and *Burmasporum*) known from inclusions in 99-million-year-old Burmese amber, Sphaerusidae has maintained its morphology and riparian habits for at least 100 million years. Thus, beetles in the genus *Sphaerius* could be considered highly specialized living fossils.

Larval examples of all myxophagan families are known, save for Lepiceridae. Myxophagan larvae are small, flattened, and possess spiracular gills. These features are all likely associated with their preference for hygropetric habitats. The dense arrangement of muscles and other structures, along with the unusual shape and size of the head, likely evolved due to their very small size. Their strong degree of miniaturization thwarted initial attempts to examine their internal organs by dissection and serial sections. Fortunately, the use of computer-generated, three-dimensional reconstructions eventually enabled researchers to develop detailed descriptions of their internal features.

← *Hydroscapha natans* (Hydroscaphidae) is a skiff beetle that inhabits algal mats along cold streams and hot springs in western North America and adjacent Mexico.

ADEPHAGA

A dephaga are an ancient group of beetles with a fossil record dating back to the early Triassic or Late Permian. The suborder includes approximately 45,000 species in the families Haliplidae, Gyrinidae, Noteridae, Meruidae, Aspidytidae, Amphizoidae, Hygrobiidae, Dytiscidae, Trachypachidae, Cicindelidae, and Carabidae.

Adephagan adults and larvae are mostly predators, although some groups are decidedly not hunters at all. Larval haliplids nibble on algae, while rhysodine carabids consume slime molds and some harpaline carabids eat seeds.

Most adephagan families occur in all major zoogeographic realms except Antarctica. However, several families are considered to have relictual distributions (meaning they were previously more widespread). For example, amphizoids, represented by only five species in the genus *Amphizoa*, are known only from western North America, central and eastern China, and North Korea. *Hygrobius*, the sole genus of Hygrobiidae, comprises six species that are patchily distributed in the western and eastern Palearctic, and across Australia. The two known species of Aspidytidae, *Aspidytes niobe*

MERU PHYLLISAE

The sole member of Meruidae, *Meru phyllisae*, is the smallest species of aquatic Adephaga. Commonly known as comb-clawed cascade beetles, they were first discovered in 1985 in Venezuela, but were not described scientifically until 2005. Few beetles have been found since and little is known of their biology. Less than a millimeter in length, these tiny, pale tan beetles inhabit the edges of a whitewater cascade that forms a slow-flowing film over exposed granite bedrock. This rock formation is part of the Guiana Shield, a region known for its biodiversity, and is home to many species found nowhere else.

↑ *Amphizoa insolens*
(Amphizoidae) ranges
from Alaska to southern

California, east to Alberta
and Wyoming; the genus
also occurs in Asia.

and *Sinaspidytes wrasei,* occur in South Africa and China, respectively.
The sole species of Meruidae, *Meru phyllisae,* occurs in southern
Venezuela. Trachypachidae is represented by only two genera. Three
of the four species of *Trachypus* live in western North America, while
the fourth species occurs in the Palearctic. Both species of
Systolosoma occur in central and southern Chile.

Supported by both adult and larval features, the monophyly of
Adephaga is undisputed. Based on habitat, adephagan beetles
may be separated into two groups, the terrestrial Geadephaga
(Trachypachidae, Cicindelidae, Carabidae) and the remaining
aquatic Hydradephaga, but recent phylogenetic studies don't
support the monophyly of either group.

POLYPHAGA

With more than 335,000 species, Polyphaga is the largest suborder of beetles. It was the first suborder recognized within Coleoptera and is currently represented by 191 families. Nearly two-thirds of all known beetles occur in just eight polyphagan families, including Staphylinidae (around 55,000 species), Curculionidae (around 51,000), Chrysomelidae (around 37,000), Cerambycidae (around 35,000), Scarabaeidae (around 32,000), Tenebrionidae (around 20,000), Buprestidae (around 15,000), and Elateridae (around 10,000). Many of the species classified in these families are well known because of their large size, horn-like armaments, spectacular colors, curious behaviors, or pest status.

WIDE RANGING

Polyphagan beetles occupy all zoogeographic realms except Antarctica and occur in both terrestrial and aquatic habitats. As the name of the suborder suggests, polyphagan beetles feed on all kinds of organic tissues, including those of plants, fungi, and animals, both living and dead. Most herbivorous species consume the roots, trunks, branches,

---

CLASSIFICATION—AN ONGOING EXPLORATION

Advanced molecular analyses of multiple genes not only help to refine our hypotheses of evolutionary relationships, but also enable us to estimate the actual age of a taxon by estimating its point of divergence with it nearest relative. These activities, coupled with the implementation of ever more sophisticated techniques for examining and elucidating fine morphological structures of both modern and fossil species, will continue to challenge our current assumptions about the classification of Coleoptera at and within the subordinal level well into the foreseeable future.

stems, leaves, flowers, fruits, and seeds of plants. Many of these species are generalists that consume a broad range of plant species, while others are more selective and specialize on plants in just one family or a single genus. Some species prefer their plant materials already decayed by fungi and other microorganisms or to have passed through the digestive systems of other animals. Several polyphagan families are largely composed of species with adults and larvae that eat only fungal tissues, including fruiting bodies and spores.

PREDATORS

Predation has evolved several times in Polyphaga. Adults and larvae of some checkered and click beetles hunt for wood-boring beetles and other insects. Predatory soldier beetles attack aphids and other sap-sucking insects, while the terrestrial larvae of fireflies prey on other beetle larvae, slugs, and earthworms. Adult ant-loving scarabs will invade the nests of select species of ants to eat their larvae, too. The larvae of several beetle families are parasitoids of the larvae of other beetles, as well as those of bees and wasps.

FAR FROM SETTLED

The classification of families within Polyphaga, as well as its relationship with other suborders, is still undergoing revision. For example, species classified in the family Jurodidae appear to have affinities with both primitive and advanced beetles. Known from multiple fossils and a single example of a living species (*Sikhotealinia zhiltzovae*) found dead in a forest cabin window in the Russian Far East, jurodids are distinguished from all other beetles by the presence of three ocelli (simple eyes) on the head. Tentatively placed in Archostemata based in part on the similarities of its thoracic structures, Jurodidae was recently moved to Polyphaga because of the similarity of their wing venation to other polyphagan species. As with its former placement in Archostemata, the affinities of jurodids within Polyphaga are uncertain.

↓ The single known specimen of *Sikhotealinia zhiltzovae* (Jurodidae) was found in the Sikhote-Alin Mountains in the Russian Far East.

EXOSKELETON

Tough, yet light in weight, the exoskeleton covers a beetle's body and protects its vital internal organs while also providing an internal scaffolding for the attachment of powerful muscles. It consists of cuticle secreted by the underlying epidermis that is one cell thick. The epidermis also contains specialized cells that form sense organs or glands that secrete various compounds on the exoskeletal surface, including a thin layer of water-repellant wax.

After molting, the soft, pale cuticle soon hardens and darkens as a result of a chemical process called sclerotization. The segments that make up the head, thorax, and abdomen are either distinct or obscure and are composed of plate-like sclerites that are sometimes delineated by cuticular membranes or narrow, groove-like sutures.

The outer surface is smooth and shiny, or variously sculpted with fine cracks, raised pebble-like tubercles, and pit-like *punctures*, the latter of which sometimes contain a single hair- or scale-like *seta*.

↓ European *Paederus riparius* and other rove beetles (Staphylinidae) have short elytra that typically expose several abdominal segments.

↓ Like many curculionids, the European willow gall weevil, *Archarius salicivorus* (Curculionidae), has a long and slender rostrum.

→ The color patterns of tiger beetles (Cicindelidae) assist with thermoregulation and also provide camouflage. Although conspicuous on the ground, the metallic red, blue, and green color patterns of *Sophiodela japonica* from Japan may help them to evade predatory birds and robber flies while they are in flight.

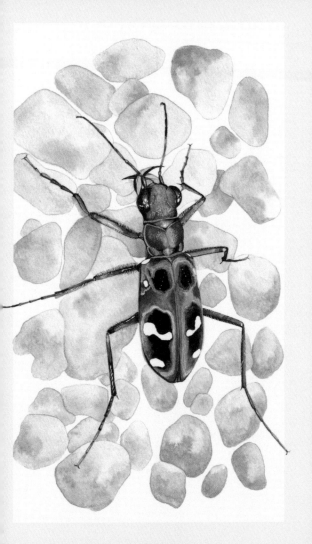

COLORS AND PATTERNS

Beetle colors are critical for mate recognition, camouflage, thermal regulation, and defense. Colors and patterns help *diurnal* species to recognize one another as potential mates. Beetles cryptically clad in somber brown and black *scales* are difficult to see against tree bark. Black desert darkling beetles quickly absorb the sun's energy in order to move about on relatively warm winter days, while their white counterparts reflect sunlight to beat the desert heat. Beetles boldly marked in contrasting patterns of black with reds or yellows warn potential predators of their chemical defenses, or mimic noxious insects so defended.

↓ The brilliantly colored golden tortoise beetle, *Charidotella sexpunctata* (Chrysomelidae), is widely distributed in the eastern United States.

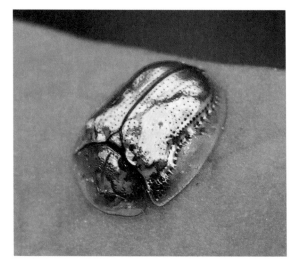

The multicolored Asian lady beetle, *Harmonia axyridis*, exhibits extraordinary intraspecific variation, as evidenced by its more than 200 different elytral color patterns. These color patterns are regulated by a single pannier gene that simultaneously promotes black pigmentation while suppressing red pigmentation. Inhibition of the pannier gene during the pupal stage results in the loss of black color patterns and normal red pattern formation. The differences in color pattern are associated with the DNA sequence of the first intron of the pannier gene, a discovery that sheds light on how intragenic chromosomal inversions can drive morphological evolution in organisms.

TINTS

Colors are either pigment-based or created by physical properties of the cuticle. Pigment-based compounds absorb or reflect various wavelengths of light. Melanin synthesized by beetles produces shades of black and brown that are more or less permanent, while colors created by food-based carotenoids (yellow, orange, red) and other pigments soon fade after death. The golden tortoise beetle, *Charidotella sexpunctata*, temporarily changes its colors from shiny red or golden orange to brilliant gold by moving pigment inside the cuticle. Its transient metallic colors are produced by light reflecting off pockets of liquid pigment within the cuticle.

SHIMMERING

Iridescent and metallic properties are produced by the exoskeleton's physical properties that scatter light. Iridescence is often created by ordered surface structures such as scales or layered nanostructures within the cuticle that reflect intense colors and shift depending on the angle of view. Beetles sporting these dazzling colors are often difficult to see in their natural environments.

HEAD

Borne on flexible, membranous necks and partially or almost completely sunken within the prothorax, beetle heads are equipped with two vital structures that help them make sense of their surroundings: compound eyes and antennae.

EYES

The sometimes kidney-shaped compound eyes are covered with multiple facetlike lenses. Each hexagonal lens sits atop a tube-shaped ommatidium that contains a cone, followed by a cluster of photoreceptor cells, or rhabdom, surrounded by pigment cells that detect light. Light funneled through the lens and cone onto the rhabdom stimulates the surrounding pigment cells and is ultimately transmitted to the brain as an electrical impulse. Each lens "sees" a portion of an entire scene. The greater number of lenses that receive light in each eye, the greater the beetle's visual acuity. Abundant light combined with light-adapted ommatidia provide diurnal beetles with good visual resolution. Nocturnal species have dark-adapted eyes that allow them to see in only low levels of available light, but this configuration sacrifices overall visual acuity. Beetles that inhabit dark caves and other subterranean habitats often have few ommatidia or lack them altogether and are completely blind.

Whirligig beetles live on the surfaces of ponds and stream pools and have compound eyes that are completely divided by a strip of cuticle called the canthus. The upper portions of their eyes are adapted

---

MALES WITH HORNS

The heads of some males, scarab beetles in particular, are adorned with spectacular horns that function as pinchers, spikes, crowbars, or scoops. These imposing structures are used to defend food sources and nesting sites attractive to females from rival males.

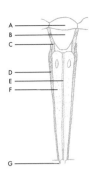

← The basic visual structure of beetles and other insects is called the ommatidium and this consists of the following:

(A) cornea, (B) crystalline cone, (C–D) pigment cells, (E) rhabdom, (F) photoreceptor cells, (G) optic nerve.

for seeing in the air, while the lower portions are adapted for seeing in water. In addition to compound eyes, some adults possess a simple eye, or ocellus, on the front of the head between the compound eyes.

ANTENNAE

The antennae are a beetle's primary organs of smell and touch. Diverse in form, they are located between the compound eyes and mandibles and usually consist of three basic parts: scape, pedicel, and flagellum. Each antenna usually has 11 articles called antennomeres. Some species possess 10 or fewer antennomeres, while a few species may have 12 antennomeres or more. Males often have longer or more elaborately modified antennae than the females, a form of sexual dimorphism. Greater antennal development in males is not only associated with increased sensory capabilities for locating females via their scent, or sex pheromone, but is also sometimes associated with guarding females while laying their eggs.

MOUTHPARTS

The chewing mouthparts of beetles usually consist of a labrum, a pair of mandibles and maxillae, and a labium. Mandibles are variously modified to cut plant and animal tissues, grind spores and pollen, or strain nutrients from various fluids. The mandibles of some species, especially in males, are greatly enlarged. Mandibles such as these are not organs of ingestion, but are used as weapons in pitched battles with rival males, and thus play a vital role in reproduction. Associated with the maxillae and labium are pairs of flexible, fingerlike palps that help beetles locate and handle food. The chewing mouthparts of weevils are often borne on a long, snout-like rostrum. Female weevils use this structure to chew a hole deep inside plant tissues, into which they can lay their eggs out of harm's way.

THORAX

The three-segmented thorax houses powerful legs and wing muscles. The first thoracic segment, or prothorax, bears the front legs and may be armed dorsally with horns and tubercles, or scooped out like a bulldozer blade. The modified forewings, or elytra, and middle legs attach to the mesothorax, while the folded membranous hindwings and hind legs are joined to the metathorax. The meso- and metathoracic segments are obscured by the elytra.

WINGS

The leathery or shell-like elytra are modified mesothoracic wings that partially or completely cover the abdomen at rest and function as stabilizers in flight. Their surfaces are variously sculpted with punctures, ridges, or grooves. Beneath them are the metathoracic hind wings. When functional, the hind wings are nearly always longer than the elytra. Hinge-like veins enable beetles to fold their hind wings beneath the elytra when not in use.

FEATHERWING BEETLES

Featherwing beetles (Ptilidae) are the smallest known beetles and are the subjects of studies on miniaturization. Their fringed and slender or paddlelike wings resemble those of similarly small wasps called fairyflies. Unlike fairyflies, ptiliids can fold and unfold their wings. They fold their wings by using special microsculptured patches on their abdominal tergites. Ptilids unfold their wings with the aid of an elastic protein called resilin, spreading curved cross-sections like retractable metal measuring tape, or by increasing blood pressure within major wing veins. Studying the patterns and mechanisms of the folding and unfolding of ptilid wings may provide clues for engineers designing miniature flying robots.

↑ Male Asian Atlas beetles, *Chalcosoma atlas* (Scarabaeidae), have cephalic and prothoracic horns, unlike the females.

LEGS

Each leg is anchored to a socket-like cavity beneath the thorax by its coxa, the first of six leg segments. The coxa articulates with a small trochanter that is usually fixed to the large and muscular femur. Following is a relatively long and slender tibia that is often modified on the front legs with rakelike extensions for digging. The tarsus, comprising up to five tarsomeres, may have adhesive or brushy pads underneath to help gain purchase on slippery surfaces, including those of food plants and the elytra of mates. Each leg terminates in the pretarsus, a segment that usually bears a pair of claws, although some dung beetles lack tarsi altogether on their front legs. In sexually dimorphic species, the legs of males and females sometimes differ in structure and/or length.

ABDOMEN

The typical beetle abdomen consists of ten segments, the last two of which are variously modified as reproductive organs and are not visible externally. A long, egg-laying tube, or *ovipositor*, is typical in females that deposit their eggs deep in soil or plant tissues. Short and stout ovipositors are usually found in species that glue their eggs to various surfaces. Male reproductive organs are often distinctive sclerotized structures that are of considerable value in species identification.

The remaining eight ring-like abdominal segments consist of a *dorsal* and *ventral* sclerite. The dorsal sclerites are called tergites. Tergites tend to be thin and flexible but are thicker and more rigid in beetles with short elytra. The posterior-most tergite is the *pygidium*. The ventral sclerites are called sternites. When visible externally, sternites are called ventrites. The breathing pores, or *spiracles*, are located along the sides of the abdomen in or near the membrane separating the tergites and ventrites.

↓ The flat, round, and brownish larvae of *Psephenus herricki* (Psephenidae) and other water pennies resemble small, segmented coins.

↓ The horn of the male European rhinoceros beetle, *Oryctes nasicornis* (Scarabaeidae), is first visible in the pupal stage.

→ A dorsal and ventral view of the North American *Derobrachus forreri* (Cerambycidae). (A) mandible, (B) compound eye, (C) antenna, (D) pronotum, (E) middle coxa, (F) elytron, (G) hind femur, (H) ventrite, (I) hind tibia, (J) hind tarsus, (K) hind claw.

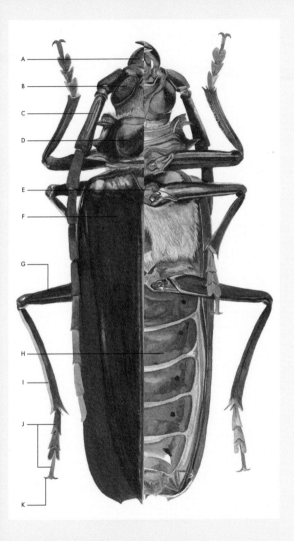

EGGS, LARVAE, AND PUPAE

B eetle eggs are usually smooth and soft, but some are distinctly sculptured and hard-shelled. The egg's shell, or chorion, is permeable to oxygen, although the eggs of some species have specialized breathing holes known as aeropyles. All eggs have micropyles, minute openings in the chorion located mostly at the end that allow sperm inside the egg for fertilization. The number of micropyles varies among and within species.

SEXLESS EATING MACHINES

Larvae have distinctive heads with chewing mouthparts for crushing, grinding, or tearing foodstuffs. Some predaceous larvae have sickle-like mouthparts to pierce and inject their prey with enzymes that liquefy the victim's tissues and organs, then use these same mouthparts like straws to suck up the fluids. Instead of compound eyes, beetle larvae usually possess one or more pairs of simple eyes called stemmata on each side of the head, although some larvae lack eyes and are blind. The antennae of most beetle larvae consist of three simple segments. Water tigers (*Hydrophilus*) use their sharp, pointed antennae in concert with their mandibles to cut up insect prey.

BEETLE PUPAE

Beetle pupae are typically of the adecticous exarate type. They lack functional mandibles (adecticous) and have legs that are not tightly appressed (exarate) to the body. Pupae often have functional abdominal muscles that allow for some movement. In some species, opposing abdominal segments have specialized teeth, or sharp edges, known as gin-traps, which defensively clamp down on the appendages of small arthropod predators such as ants and mites.

The larval thorax consists of three very similar segments, the first of which may have a thickened dorsal plate. Legs, if present, have six or fewer segments, but are often greatly reduced or absent in species that feed inside plant tissues or parasitize other insects.

Most larvae have nine- or ten-segmented abdomens. Soft and pliable abdomens allow rapid expansion as the larva feeds and reduces the number of times they have to shed their exoskeleton in order to increase their capacity. Some legless terrestrial larvae possess fleshy, wartlike protuberances on their abdomens that help provide traction as they move about in soil or wood. Aquatic larvae may have simple or branched abdominal gills laterally or ventrally. The abdomens of many beetle larvae terminate in a pair of short, fixed, or segmented extensions called urogomphi.

LARVAL FORMS

Larval beetles are grouped together based on their body form. The eruciform larvae of lady beetles and some leaf beetles are slow and caterpillar-like in appearance. The plump, C-shaped scarabaeiform grubs of scarab beetles and their relatives have distinct heads and well-developed legs suited for burrowing in soil and rotten wood. In spite of having well-developed legs, the larvae of flower chafers (Scarabaeidae) generally crawl on their backs. Larval click beetles, as well as those of many darkling beetles, are elateriform and have long, slender bodies with short legs and tough exoskeletons. Stout, legless weevil grubs are considered vermiform because of their maggot-like appearance. In contrast, campodeiform larvae of ground, whirligig, predaceous diving, water scavenger, and rove beetles are long, flat, and leggy. The cheloniform water penny larvae are broadly oval, distinctly segmented, and turtlelike, while the pillbug-like larvae of some carrion beetles are referred to as onisciform. Fusiform beetle larvae are broad across the middle and tapered at each end.

→ Aquatic *Hydrophilus* (Hydrophilidae) larvae are voracious predators of insects and other invertebrates, as well as tadpoles and small fish.

HOLOMETABOLY

B eetles, butterflies and moths, ants, bees, wasps, flies, fleas, and their relatives all develop by holometaboly. Commonly known as complete metamorphosis, holometabolous development is characterized by four distinct developmental stages: egg, larva, pupa, and adult. Each of these stages is adapted to a particular suite of environmental factors that enhances the overall chances of a beetle surviving and thriving, especially in temperate climates. Most adults and their larvae lead largely separate lives, thus limiting interspecific competition for food and space.

Although considered one of the key evolutionary innovations that led to the incredible diversity of beetles and other insects, there is currently little consensus as to how or why holometaboly evolved. While the idea that holometaboly allows a single species to occupy multiple niches and utilize different resources is certainly appealing, it reveals little about the selective mechanisms that led to its evolution, including the unique development of the relatively inactive and purely transformational pupa.

↓ The pupae of most beetles are adecticous (lacking functional mandibles) and exarate, meaning that the appendages are free.

↓ Most beetles reproduce sexually and must mate before they can produce and lay viable eggs.

→ European stag beetles, *Lucanus cervus* (Lucanidae), lay eggs in decaying wood. The larvae pass through three instars before pupating within a case made of soil and coated inside with gut secretions. Adults live nearly a year or longer in captivity.

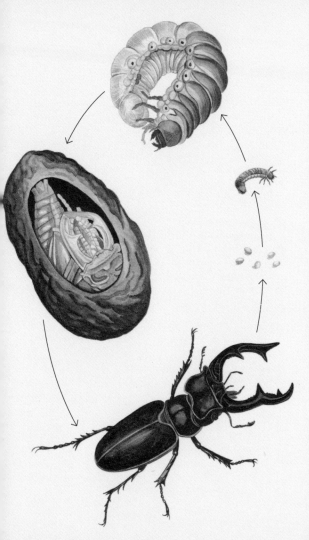

GROWTH

Upon hatching, most beetle larvae begin feeding almost immediately and grow rapidly. As with all insects, growth and development is driven by complex hormonal interactions regulated by an endocrine system composed of neurosecretory cells within the central nervous system and specialized endocrine glands. High levels of juvenile hormone released into the hemolymph by the corpora allata inhibit the development of adult structures, while increased amounts of ecdysteroids produced by the prothoracic glands induce molting and stimulate growth.

MOLTING

Working in concert, fluctuating levels of juvenile hormones and ecdysteroids trigger the reorganization of tissues and the shedding of the exoskeleton, a process known as molting. The outgrown exoskeleton is replaced with a new and roomier version secreted by an underlying layer of epidermal cells. The stage between each larval molt is referred to as an instar. Most species pass through a definite

---

HYPERMETAMORPHOSIS

Typically, each successive instar resembles a larger form of its predecessor. However, parasitic species undergo a special type of holometaboly called *hypermetamorphosis*. For example, the larvae of cicada parasite beetles, blister beetles, and wedge-shaped beetles are all characterized by having two or more distinct larval forms. The active and leggy first instar, or *triungulin*, is adapted for seeking out the appropriate host. Once a host has been located, the triungulin molts into a more sedentary larval form with short, thick legs. It goes on to metamorphose into a fat, legless grub that eventually develops into a more active short-legged grub that spends most of its time preparing a pupal chamber.

---

number of instars, usually ranging from three to five. Hister beetle larvae have as few as two instars, while those of some skin beetles may have as many as seven. In western North America, rain beetles (*Pleocoma*) may undergo more than a dozen instars.

PUPATION

Concurrent with a reduction of juvenile hormone at the onset of pupation, increased levels of ecdysteroids stimulate the beginning of structural differentiation. Larval tissues and organs once dedicated to eating and growing are restructured for locating mates and reproduction. Lacking functional mouthparts and appendages, beetle pupae are non-feeding and mostly sessile. However, some species do have the ability to flex their abdomens just a bit,

↑ Larviform females of *Dulticola* (Lycidae) are commonly known as trilobite beetles because of their resemblance to the long extinct marine arthropods.

sometimes as a limited means of defense. Consensus on the evolutionary origins of the pupal stage has yet to be achieved. Some researchers suggest that the pupa is derived from a final larval stage, while others believe it is a modified preadult.

Glowworms and some fireflies undergo a modified pupal stage. In these species, adult females closely resemble the last larval instar and are referred to as *larviform*. Larviform females typically have reduced elytra, if they have them at all. These wingless adults are best distinguished from larvae externally by the presence of compound eyes and internally by their reproductive organs.

ECLOSION

The requisite combination of time, temperature, and precipitation triggers adult emergence from the pupa (known as eclosion). At first soft and pale, the exoskeletons of these *teneral* adults soon begin to harden and darken as they undergo sclerotization, a chemical process akin to the tanning of leather. Fully developed adult beetles will never molt again and are soon ready to mate and reproduce.

DIGESTION

The digestive tracts of beetles are variously adapted for ingesting and processing an astonishing variety of foods. Species that consume nutrient-rich animal tissues don't need the long and convoluted gut required by herbivorous beetles to extract sufficient nutrition solely from plant materials. Like those of most insects, the digestive tract of beetles is divided into three functional regions: foregut, midgut, and hindgut.

FOREGUT: INTAKE

The foregut is the site of ingestion, mechanical breakdown, and temporary storage of food. Food cut and ground or strained by the mandibles is drawn into the esophagus by powerful cibarial and pharyngeal muscles, sometimes with the aid of saliva. The esophagus usually extends to an expanded storage chamber, the crop. Behind the crop is the proventriculus, a valve that controls food moving into the midgut and, in predatory ground beetles and plant-feeding weevils, serves as a food-grinding organ.

MIDGUT: ABSORPTION

The midgut produces and secretes digestive enzymes, absorbs nutrients, and transports remaining food and waste to the hindgut. Inside the abdominal cavity are worm-like Malpighian tubules that are involved in excretion and osmoregulation. These blind-ended excretory organs originate near the junction of the midgut and hindgut and extract water and waste from the hemolymph. Some beetles rely on symbiotic microorganisms living in the midgut to assist them with digestion.

HINDGUT: WASTE MANAGEMENT

The remaining contents of the midgut pass through the pyloric valve before entering the hindgut. Here water, salts, and other minerals are absorbed prior to elimination of feces through the rectum and out the anus.

THE ROLE OF SYMBIOTIC BACTERIA

In beetles and other holometabolous insects, symbiotic bacteria are either fully integrated with their morphology or physiology (obligate symbionts) or not (facultative symbionts) and are not utilized just as digestive aids. Wood-feeding longhorn beetle and bark beetle larvae depend on nitrogen-fixing symbiotic bacteria to make up for the low nutritional quality of their carbohydrate-rich diets. Other species depend on their bacterial symbionts to help manage and suppress potentially dangerous toxins associated with their foods. Coffee berry borers (*Hypothenemus hampei*) rely on facultative gut bacteria to break down harmful caffeine, while those of cowpea beetles (*Callosobruchus maculatus*) help them to detoxify pesticides.

↓ This predatory European golden ground beetle, *Carabus auratus* (Carabidae), is in the process of attacking an earthworm.

SENSES

Beetles sense the world around them with the aid of highly sensitive receptors borne on various surface structures, especially on their heads. The complex nervous system of beetles integrates external stimuli with internal physiological information to generate a diverse array of behaviors.

BASIC NEUROLOGY

The nervous system of beetles consists of the visceral nervous system, peripheral nervous system, and central nervous system (CNS). The visceral nervous system innervates the gut, endocrine and reproductive organs, and the tracheal system. The peripheral nervous system connects the muscles with the CNS and visceral nervous system, as well as the cuticular sensory structures that receive visual, chemical, tactile, and thermal stimuli from the beetle's immediate environment.

The CNS is the primary division of the nervous system and consists of a series of bundled nerves, or ganglia, connected by ventral paired nerve chords called connectives. Inside the head are the brain and suboesophageal ganglion. The brain innervates the primary sense organs (eyes, antennae) and receives signals from elsewhere in the body, while the suboesophageal ganglion supplies nerves to the mouthparts.

MOSTLY DEAF

The ability to hear is rare among Coleoptera, but some tiger beetles and dynastine scarabs have earlike structures capable of hearing ultrasounds, including frequencies used by echolocating bats. Tiger beetles (Cicindelidae) have a pair of domed tympanic membranes atop their first abdominal segment, while rice beetle (Scarabaeidae) "ears" are located on the neck membrane just behind their heads.

EYES

Most beetles have a pair of well-developed compound eyes to help them navigate their environments and avoid danger. Dung-rolling scarab beetles use the sun, moon, and the Milky Way as visual cues to help keep them on a straight path. The bulbous eyes of diurnal tiger beetles affords them binocular vision that helps them gauge the distance of prey. The arrangement of simple eyes, or stemmata, not only affords both far and near vision to larval sunburst diving beetles (*Thermonectus marmoratus*), but it may also give them depth perception.

↑ The fan-like surfaces of male *Polyphylla* antennae possess many structures for sensing odor released by females.

ANTENNAE

Primarily organs of smell and touch, the antennae of beetles are paired appendages possessing special receptors for finding food and detecting pheromones. Males often have longer or more elaborate antennal structures resembling fans or feathers, modifications that increase the surface area available for sensory structures. The antennae of some ground beetles and rove beetles sometimes have special comb-like structures used for grooming their legs and feet. During courtship, male oil beetles use their antennae to grasp those of their partner. Whirligig beetles use their antennae to detect ripples generated by insect prey trapped on the water surface, but claims of their ability to navigate by echolocation remain unsubstantiated.

MOUTHPARTS

The fingerlike palps associated with a beetle's maxillae and labium serve primarily as touch and taste receptors. The tips of these paired structures are usually membranous in both adults and larvae and bear short sensilla thought to be associated with taste. In some beetles, these structures are greatly modified, suggesting that they might have additional sensory functions. Snail-eating *Cychrus* ground beetles on the hunt bring their elongated palps tipped with enlarged, spoon-shaped flanges in contact with the soil, presumably to detect their prey's slime trails.

LOCOMOTION

T he mobility and strength of beetles is driven by antagonistic pairs of internal muscles fused with and working against rigid exoskeletal segments. Adults have stiff cuticular exoskeletons, while their soft-bodied larvae rely on a hydrostatic skeleton made turgid by thousands of criss-crossed body wall muscles pressing against a fixed volume of hemolymph within the body cavity.

HERCULEAN FEATS

Beetles can perform extraordinary feats of strength for their size. For example, the male horned dung beetle *Onthophagus taurus*, which measures only ²/5 in (10 mm) in length, can pull up to 1,141 times its own body weight. This is equivalent to an average-sized male lifting six double-decker buses packed with commuters. Male dung beetles use their power to push rival males out of underground tunnels occupied by a female.

AQUATIC BEETLES

Locomotion in aquatic beetles is dominated by the forces of drag, buoyancy, viscosity, and surface tension. While most species move in aquatic environments by swimming through the water or crawling on submerged plants and other substrates, some rely on the cohesive forces of water molecules at the water–air interface to get about. Whirligig beetles are the best known for their utilization of the water's surface. Recently, small hydrophiloid beetles have been observed crawling and resting on the underside of the water's surface. An understanding of the mechanisms enabling this activity could assist in the development of bio-inspired aquatic adhesives and robotics.

AGILITY

The legs and wings are powered by muscles that are concentrated within the thorax. By relaxing and contracting muscle pairs associated with variously modified legs, beetles can walk, run, climb, burrow, or swim as they search for food and mates. While these methods of getting about work well for covering short distances, the power of flight enhances their abilities to disperse over larger areas.

AIRBORNE

Before taking flight, beetles lift their elytra to free the membranous flight wings folded underneath. The flight wings quickly expand as the blood pressure within their network of veins increases. Once airborne the elytra act as stabilizers, while the flight wings function as airfoils that simultaneously provide beetles with maneuverability and the prerequisite aerodynamic qualities to keep them aloft.

← A glorious jewel scarab, *Chrysina gloriosa* (Scarabaeidae), from the American Southwest and adjacent Mexico, takes to the air.

GAS EXCHANGE

The exchange of the respiratory gases oxygen and carbon dioxide is facilitated by an internal system of branching, air-filled tubes called trachea. Extending throughout the body, the tracheal system directly supplies internal organs and tissues with oxygen and removes carbon dioxide. Oxygen enters the tracheal system through valved spiracles that are openings located along the sides of the body. Carbon dioxide generated by cellular metabolic processes exits the body via the trachea and spiracles.

In aquatic beetles, gas exchange is accomplished by keeping the spiracles in contact with the air. Hydrophilids (water scavenger beetles) and dytiscids (predaecous diving beetles) regularly surface to replenish bubbles trapped beneath their bodies or elytra, respectively. Sedentary grazers like riffle beetles (Elmidae) and long-toed water beetles (Dryopidae) that inhabit shallow, well-oxygenated waters rely on *plastron respiration*. Clothed in dense, velvety, and water-repellant pubescence called a *hydrofuge*, their bodies are permanently enveloped by a thin layer of air called the plastron. Dissolved oxygen from the surrounding water steadily diffuses into the plastron as carbon dioxide diffuses out.

← A bubble temporarily exposed at the elytral apices of a predaceous diving beetle (Dytiscidae) draws in dissolved oxygen from the water.

↙ The abdominal spiracles of the larvae of the coconut rhinoceros beetle, *Oryctes rhinoceros* (Scarabaeidae), are clearly visible along the sides of their abdomens.

→ Using micro-computed tomography, the tracheal system of the adult mealworm, *Tenebrio molitor* (Tenebrionidae), can be visualized and grouped into functional areas or modules. The cephalon-prothoracic module is represented in orange, while the thoracic and abdominal modules are shown in red and olive green, respectively.

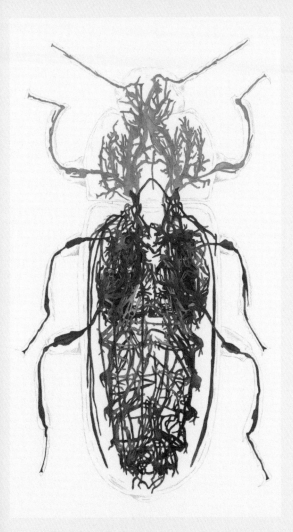

REPRODUCTIVE SYSTEM

Most beetles must reproduce sexually. The basic structure and function of the reproductive systems of beetles resemble those of vertebrates. The male's testes produce sperm inside packets called spermatophores, while the female's ovaries produce eggs. During copulation, males deposit their spermatophores directly inside the female by inserting their penis-like reproductive organs into her genital tract. These ornately shaped organs are thought to correspond specifically with the vagina of females of the same species. This unique fit, combined with species-specific pheromones and behaviors, works like a lock and key and is thought to prevent beetles from mating with the wrong species.

In addition to sperm, the spermatophore also contains other vital compounds and nutrients. Females temporarily store sperm in a special gland called the *spermatheca*. Sperm released from the spermatheca fertilize the mature eggs as they pass down from the ovaries and out the body through the egg-laying tube, or ovipositor.

↓ The structure of the male's reproductive organs, especially the parameres, are often considered diagnostic and are useful for distinguishing closely related beetle species. Shown here are three different views of the parameres of *Pentodon idiota* (Scarabaeidae).

→ Male and female Hercules beetles, *Dynastes hercules* (Scarabaeidae), preparing to mate. One of the most recognizable beetles in the world, this species is found from Mexico southward to northern South America, and also on several Caribbean islands.

COURTSHIP AND MATING

Most beetles reproduce sexually and invest enormous amounts of time and energy in finding mates and copulating. They engage in various species-specific behaviors that have evolved to help synchronize the sexes in both time and place, including *bioluminescence*, intrasexual competition, and pheromones.

LIGHTS ON

Bioluminescence in fireflies serves as both a feeding deterrent and a form of sexual communication. Present in all life stages (egg, larva, pupa, adult), bioluminescence warns educated predators of their bitter taste. Bioluminescent adults also use their lights as a means of sexual communication. The light signal of each species is unique and expressed as a continuous glow or as a series of discrete and precisely timed flashes.

Firefly bioluminescence results from a chemical reaction involving calcium, adenosine triphosphate, luciferin, and the enzyme luciferase in the presence of oxygen. Adenosine triphosphate (ATP) provides energy. Oxygen reaches special light-producing cells called photocytes via the trachea. Controlled by the nervous system, the amount of oxygen, along with nitric oxide, octopamine, and hydrogen peroxide, affects the color, brightness, and duration of the light. Unlike that of an incandescent light bulb, very little of the energy in this reaction is lost as heat, thus a firefly's light is cold.

FEMALE CHOICE

Previous studies on insect mating systems have focused mostly on how males acquire mates. Male Japanese rhinoceros beetles, *Trypoxylus dichotomus septentrionalis*, certainly engage in battles, but recent observations revealing their stridulatory behavior and production of cuticular hydrocarbons suggest that females may have the final say by choosing males based on their sounds and smells.

IS BIGGER BETTER?

Large male beetles armed with elaborate horns are textbook examples of the resource defense system, a strategy illustrated by males in the genera *Dynastes* and *Megasoma*. These beetles aggressively use their head and prothoracic armaments against rival males, especially when guarding naturally occurring sapping tree wounds that are attractive to females. In these bloodless battles, the larger males are more likely to repel their lesser endowed competitors and mate with nearby females. However, less powerful males can still mate successfully by avoiding confrontations with larger males.

↑ Samurai helmets were inspired by the appearance of male Japanese rhinoceros beetles, *Trypoxylus dichotomus* (Scarabaeidae).

GOOD SCENTS

Pheromones are employed by many beetles to attract and locate mates over long distances. Species that use pheromone-based mating systems are often sexually dimorphic, with males having more elaborate antennae than females. The increased surface areas of these antennae provide ample space for special receptors that can detect just a few molecules of the female's pheromone. Once detected, males take to the air and fly in a zigzag pattern until they locate the female's odor trail, then follow the ever-increasing concentration of pheromone molecules to its source. "Calling" females can be tracked from considerable distances, even when they are hidden among tangles of vegetation and leaf litter, or at the entrances of their burrows.

Copulation usually commences upon contact, but some beetles first engage in courtship behaviors that induce sexual receptivity. For example, female oil beetles and some lepturine longhorn beetles release cuticular hydrocarbons from their elytra that stimulate males to stroke them with their antennae, mouthparts, legs, and genitalia. These stereotypical behaviors help these beetles to recognize one another as suitable partners.

PARENTAL CARE

Carrion, dung, bark, and other beetles are known to engage in elaborate nesting and food preparation behaviors to ensure the survival of their offspring. Some Neotropical tortoise beetles use their shield-like bodies to protect their eggs and larvae. However, parental care in most beetles is limited to depositing their eggs singly or in batches on or near suitable larval foods.

EGG LAYING

Plant-feeding species drop their eggs at the base of larval food plants or glue them to vegetation. Longhorn beetles lay their eggs in cracks, crevices, and wounds in bark. Aquatic beetles affix their eggs to plants, rocks, chunks of wood, and other submerged objects. Ground-dwelling scavengers oviposit in leaf litter, compost, dung, decaying logs, carrion, and other accumulations of decaying organic matter.

EVOLUTION OF THE
PARENTAL-LARVAL RELATIONSHIP

Parental care is universal among species of *Nicrophorus* carrion beetles that bury small vertebrate carcasses as food for their larvae. By manipulating pre- and post-hatching care levels, researchers observed considerable variation between species in terms of larval dependency on parental feeding. The degree of parental care may have been driven by defense of the carcass and offspring that eventually led to the extended care afforded by one or both parents to their offspring, while the level of offspring dependance was likely driven by a coadaptation of adult and larval behaviors. Comparative studies taking into account the phylogenetic relationships of burying beetles may lead to further insights into the evolution of both adult and larval behaviors.

↑ Female tortoise beetles, *Acromis sparsa* (Chrysomelidae), can defend their larvae more effectively against insects than larval parasitoids.

FOOD AND SHELTER

Some female beetles exhibit interesting egg-laying behaviors. For example, leaf-mining jewel beetles and weevils insert their eggs between the upper and lower surfaces of leaves, thus providing both food and shelter for their larvae. Certain ground beetles lay their eggs in cells constructed of mud and bits of vegetation, while leaf beetles encase eggs in their own feces laced with distasteful chemical compounds to deter predators. Leaf-rolling weevils lay their eggs on a leaf, then roll the leaf into a barrel-shaped nest using their legs and mandibles. Longhorn beetles known as girdlers chew a ring around a living branch before laying a single egg on the outer tip. The girdled branch soon dies and falls to the ground with the larva feeding and developing inside.

LARVAL BEHAVIOR

Beetles spend most of their lives as larvae buried in soil and humus, hidden among litter and fungus, or tunneling through wood and various other plant tissues. However, the larvae of some leaf beetles feed conspicuously on leaves and camouflage themselves with their own exuviae and feces. Plant-feeders and scavengers are sometimes considered pests when they attack crops, damage trees managed for timber, or infest stored products.

The larvae of Adephaga are mostly predators that attack other insects and invertebrates. Larval cicada parasite, blister, and wedge-shaped beetles are parasitoids, which are highly specialized predators that attack the eggs and larvae of specific insect species.

Once they reach maturity, most beetle larvae seek a place to construct a chamber in which to pupate. Aquatic species leave the water to pupate under rocks and logs nearby. Wood-boring species tunnel toward the surface and usually pupate just underneath the bark. Some rove beetle and weevil larvae pupate within a loosely woven silken cocoon.

↓ Striped blister beetles, *Epicauta vittata* (Meloidae), develop by hypermetamorphosis. Their first instar larvae are called triungulins.

↓ The plump larvae of Colorado potato beetles, *Leptinotarsa decemlineata*, feed externally on the leaves of solanaceous plants.

→ Lunging backward from their burrows, tiger beetle (Cicindelidae) larvae seize insect prey with their powerful mandibles. The hooks on the back make it difficult for predators to pull the larva out of its burrow.

DEFENSE STRATEGIES

Beetles are variously modified physically to defend themselves against predators. For example, with the aid of adhesive pads under their tarsi, tortoise beetles will hunker down on a leaf to shield themselves from the prying mandibles of ants. Tiger beetles simply outrun their enemies. Click beetles flip themselves up into the air with an audible click. For larger beetles, size alone, backed up by strong mandibles, pincher-like horns, and powerful legs tipped with sharp claws, will deter nearly all but the hungriest of predators.

PLAYING POSSUM

Thanatosis, or death feigning, is employed by hide beetles, some darkling beetles, zopherid beetles, weevils, and others. Some Hybosoridae and Leiodidae can partly or completely roll their bodies up. Motionless and with their appendages carefully tucked away, these beetles strongly resemble seeds or bits of disarticulated insects.

CHEMICAL DEFENSES

Ground and predaceous diving beetles possess specialized thoracic and abdominal organs that produce toxic combinations of aldehydes, esters, hydrocarbons, phenols, and quinones laced with various acids. Some beetles engage in *reflex bleeding*, a behavior where powerful chemicals in the blood-like hemolymph are released from their leg joints as feeding deterrents. Lady beetles release bitter-smelling and -tasting alkaloids, while blister beetles exude caustic cantharidin.

APPEASEMENT

Ant-loving beetles in several families often have trichomes, tufts of specialized hair-like structures on the thorax or abdomen, which are associated with specialized glands that produce appeasement substances. Trichomes likely function as wicks that dispense substances that attract ants and tone down their aggressive behaviors, thus protecting these highly specialized myrmecophilous beetles from harm and facilitating their integration into the host colony.

BOMBARDIER DEFENSE

The defensive glands of bombardier beetles act as binary chemical weapons. Rather than storing benzoquinone, the beetles synthesize this noxious chemical internally by injecting the contents of a pair of two-chambered glands into a

↑ The Asian bombardier beetle, *Pheropsophus jessoensis* (Carabidae), audibly defends itself with a cloud of noxious, boiling hot gas.

common chamber, and exploding it from their anus with a pop. Although the evolutionary history of the bombardier beetle's defense mechanism is still unknown, evolutionary biologists hypothesize that it likely evolved from the defensive pygidial glands found in all carabid beetles via a series of incremental steps. Proponents of pseudoscientific creationism and intelligent design often refer to bombardier beetles and erroneously characterize their unique defense mechanism as too complex to have evolved by natural selection.

CRYPSIS, MIMICRY, AND MIMESIS

The evolution of defensive colors and patterns, along with various surface structures and behaviors, was likely driven by birds and other diurnal predators. Cryptic beetles possess colors and patterns that render them well camouflaged against specific backgrounds such as sand, leaves, and bark. Somber-colored longhorn beetles and fungus beetles, cryptically mottled in various shades of browns, grays, and greens, almost disappear among the lichen-encrusted bark of trees.

Recent studies suggest that iridescence in tiger beetles is also a form of crypsis. Their elytra are packed with minute surface structures that reflect different qualities of light to create dull browns and greens that camouflage them in their surroundings.

WARNING SIGNS

Soldier beetles, fireflies, net-winged beetles, and blister beetles all defend themselves with various noxious chemicals extracted from their host plants or synthesized by special glands in their bodies. These sluggish insects warn potential predators of their foul odor and taste by sporting black bodies often conspicuously clad with patterns of red, yellow, or orange. Such bold markings, known as *aposematic* or warning colors, simultaneously repel experienced predators while quickly bringing novices up to speed. Bioluminescence in all life stages of fireflies, as well as iridescence in many other beetles, are also thought to be forms of aposematism. Sudden flashes of iridescence in tiger beetles and jewel beetles, along with bold eyespots in some click beetles and scarabs, may confuse or startle predators momentarily, but such hypotheses need testing.

↓ The cryptically marked New Zealand fungus weevil *Sharpius venustus* (Anthribidae) is named after British coleopterist David Sharp (1840–1922).

IMPERSONATORS

Lacking their own defenses, some beetles imitate the appearance and behavior of noxious insects, an adaptation known as Batesian mimicry. The bodies of various flower-visiting beetles, such as those of boldly colored jewel beetles and longhorn beetles, or fuzzy-bodied scarabs, strongly resemble stinging wasps and bees, respectively. These impostors drive home the ruse with their highly animated movements. Boldly marked *Enoclerus* checkered beetles running along tree limbs as they search for bark beetle prey strongly resemble velvet ants, wingless wasps that are infamous for their painful stings.

Stinging insects are not the only models for beetles seeking protection. Some click beetles in the Eurasian and North American genus *Denticollis*, along with various other longhorn beetles, moths, and cockroaches, mimic distasteful fireflies, soldier, and net-winged beetles.

Müllerian mimicry involves two or more defended species inhabiting the same region that share similar aposematic markings. Predators quickly learn to avoid boldly marked individuals, thus protecting all species that are similar in appearance. Best known among butterflies, Müllerian mimicry complexes involving various groups of boldly marked insects appear to be modeled after unpalatable fireflies, soldier beetle, and net-winged beetles.

FAKING IT

Mimesis is the resemblance of an organism to inanimate or neutral objects from the point of view of a predator. Small and chunky warty leaf beetles strongly resemble unappetizing caterpillar feces or seeds. Some beetles possess disruptive color patterns and/or highly reflective surfaces that make them look decidedly less beetle-like to predators. For example, the dorsal surface of the mostly black *Heilipus squamosus* from the southeastern United States has irregular white markings that create a pattern more suggestive of a bird dropping than that of a weevil.

FUNGUS AND BEETLES

Fungus-feeding beetles, collectively known as fungivores or mycophages, have evolved multiple times in Coleoptera. Pleasing fungus beetles, handsome fungus beetles, forked fungus beetles, and fungus weevils feast on mushrooms, puffballs, and bracket fungi. However, beetles that feed on these conspicuous fruiting fungal bodies represent only one aspect of the complex relationships between beetles and fungi. Yeasts, molds, mildews, and mycorrhizal fungi found on plant surfaces and among roots also serve as food for food for fungivorous coleopterans..

COZY RELATIONS

Many beetle–plant interactions are possible because of a beetle's intimate association with fungi. Beetles in the weevil subfamily Scolytinae are completely reliant upon their mutualistic relationships with plant pathogenic fungi. As they chew intricate tunnels beneath the bark or, less commonly, into the sapwood, females inoculate the walls of their galleries with fungal spores and hyphae stored within *mycangia* (specialized structures on their bodies that sustain the fungi) as they move from tree to tree.

BARK AND AMBROSIA BEETLES

Scolytines are divided into two major feeding guilds: bark beetles and ambrosia beetles. Bark beetles tend to be more specialized in their host tree preferences and rely on their fungal symbionts to disrupt the host tree's ability to defend itself with sap by blocking its resin canals with hyphae. With the tree's defenses disabled, bark beetles and their larvae are free to feed on the phloem and sapwood unfettered. In contrast, ambrosia beetles are more generalists in terms of the trees they colonize and utilize their symbiotic fungi primarily as food for both themselves and their brood.

CYPHEROTYLUS CALIFORNICUS

This dull black beetle has blue or sometimes purplish elytra with scattered black dots. It commonly occurs in North America, from Wyoming and Kansas south to Mexico. In summer, the adults and black, spiny larvae graze on soft polypore fungi growing on living and dead oaks, dead conifers, and other trees, especially in relatively shady and moist habitats. Eggs are laid on or near fungi. Mature larvae hang upside down in groups and pupate while still attached to shed their larval exoskeleton. Adults live up to three months and their elytral colors soon fade to a dull pale yellow after death.

↑ *Cypherotylus californicus* (Erotylidae) belongs to a family of insects commonly known as pleasing fungus beetles.

POLLINATORS

Using phylogenetic techniques incorporating both fossils and representatives of extant taxa, coleopterists have developed hypotheses as to how herbivory in beetles evolved. Modern lineages of beetles associated with conifer and cycad pollen are hypothesized to have first appeared early in the Jurassic, long before the appearance of bees or butterflies. These beetles were likely among the first pollinators of cone-bearing conifers and their allies (gymnosperms) and of early-flowering plants (angiosperms). Careful examination of fossil evidence suggests that specialized feeding habits focused on pollen and other living plant tissues likely developed from more generalized feeding habits involving the consumption of fungus or decayed plant and animal tissues.

MONOCOTS

Cycads, an ancient group of gymnosperms, are pollinated primarily today by select species of beetles in the families Boganiidae, Chrysomelidae, Curculionidae, and Nitidulidae. Beetle pollination is prevalent among plants in several families of primitive angiosperms, too, including the monocotyledonous palms, aroids, and water lilies. Weevils and sap beetles are among the top coleopteran pollinators of palms, while aroids and water lilies are reliant on, at least partially, cyclocephaline scarabs. Flowers of the Neotropical aroids *Dieffenbachia* and *Philodendron* use their pungent scents and metabolic heat to attract scarabs in the genus *Cyclocephala* to their inflorescences, as do Amazonian giant water lilies (*Victoria amazonica*). All of these flowers imprison their beetle visitors until the next evening when they are

→ The flowers of the giant water lily (*Victoria amazonica*) produce a fruity odor that attracts its scarab beetle pollinators.

released with their bodies covered in pollen. The African Nile lily (*Nymphaea lotus*) attracts and briefly confines the Old World's only cyclocephaline scarab, *Ruteloryctes morio*, in a similar fashion, although these nocturnal beetles are not such efficient pollinators of the flowers as are several species of bees that visit the flowers just after dawn.

DICOTS

Primitive flowering dicots also rely on beetle pollination. Various magnolia flowers use heat and yeasty odors to attract flower scarabs, softwing flower beetles, sap beetles, tumbling flower beetles, and weevils, among others. Of the 2,400 species of plants in the custard apple family (Annonaceae) found in both the Old and New World Tropics, about 90 percent are thought to be pollinated by beetles. The light-colored flowers lure beetles with their fruity odors. These flowers are attractive to beetles not only as sources of pollen, but also because their structures provide shelter and a place to find mates.

Many species of Scarabaeidae, Buprestidae, Cantharidae, Lycidae, Meloidae, Mordellidae, Cerambycidae, and Curculionidae regularly visit all kinds of flowers. Most of these and other flower-visiting beetles simply eat and defecate their way through flowers and are not particularly effective pollinators. As a result, they are typically dismissed by pollination biologists as "mess and soil pollinators." However, careful examination of their mouthparts and other morphological features in relation to their feeding preferences and flower-visiting behaviors are needed to determine whether or not they are truly pollinators.

The pollen-feeding flower chafer *Trichostetha fascicularis* and various monkey beetles inhabiting southern Africa possess mouthparts specifically adapted for dealing with pollen. Dense setal brushes on their maxillae sweep pollen grains into the mouth where they are pulverized by specialized mandibles. These and other fuzzy, bee-like scarabaeoid beetles that regularly visit flowers, including the European *Amphicoma*, are likely pollinators, but their specific ecological roles are in need of further study.

DUNG BEETLES

Dung is a nutrient-rich, yet ephemeral resource and several families of beetles contain species that regularly use it as food for themselves and their young. Dung scarabs have evolved a striking array of behaviors with regards to dung handling, nest construction, and brood care. Based on their strategies for securing and utilizing animal feces, dung beetles are divided into three basic groups: endocoprids, paracoprids, and telecoprids. Endocoprids tunnel directly into the dung, while paracoprids dig tunnels beneath or beside the pile. Telocoprids fashion small chunks of the stuff into balls for their own sustenance, or as brood balls into which an egg is inserted and then rolled away and buried. All of these activities contribute to the physical breakdown of feces and recycle their nutrients into the soil. Fossil and molecular evidence suggests that these dung-handling behaviors may have evolved with the dinosaurs during the Lower Cretaceous (115–130 Mya).

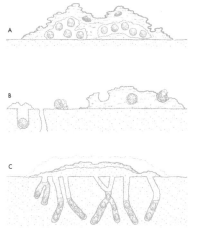

A

B

C

← The nesting behaviors of dung beetles are broadly categorized into three groups based on the relative positions of the beetles' tunnels with the food source, as follows: (A) endocoprid, (B) telocoprid, (C) paracoprid.

→ *Kheper nigroaeneus* (Scarabaeidae) inhabits the eastern savannas of southern Africa. Males construct brood balls using the dung of elephants, buffalo, rhinoceroses, cattle, and other animals, then release pheromones to attract a female.

CARRION BEETLES

Like dung, carrion is also a valuable, yet ephemeral resource for beetles that requires decaying animal flesh to sustain themselves and their young. After death, carcasses are soon beset by all kinds of necrophagous insects, including various kinds of beetles and microorganisms. To avoid this intense competition, red and black burying beetle (*Nicrophorus*) males and females work in pairs to quickly bury small carcasses as food for their young. Once buried, the fur or feathers of the dead animal are removed and the remains carefully reshaped and rolled into a ball. The beetles lick the remains

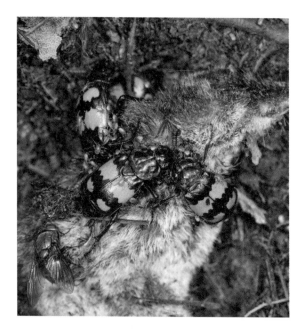

to coat them in saliva laced with antimicrobials that slow decay. Female burying beetles deposit their eggs on the walls of the burial chamber, where they will usually remain until their larvae complete their development. Upon hatching, the young larvae dine on droplets of digested carrion regurgitated by their parents into a broad depression on the surface of the remains. Soon, the larvae will grow large enough to begin feeding directly on the carcass.

STRIDULATION

In a behavior known as stridulation, *Nicrophorus* and other beetles are known to be able to produce audible squeaking sounds by moving a scraper-like plectrum on the elytral apices across paired file-like pars stridens on the abdomen. These beetles *stridulate* during courtship, carcass preparation, and parenting. The notion that stridulation is used to communicate with larvae has recently been called into question. The presence of antennal olfactory structures and stridulatory files on beetles preserved in Burmese amber suggests that modern burying beetle behavior was likely established before the Middle Cretaceous.

← A group of common sexton beetles, *Nicrophorus vespillo* (Staphylinidae), cooperate in order to bury and prepare a dead shrew as food for their future brood.

PREDATORS, PARASITES, AND INQUILINES

Predaceous beetles hunt mostly for insects and other small arthropods and invertebrates. However, the relatively large and powerful adults and larvae of *Dytiscus* are capable of capturing small vertebrates such as fish and amphibians. Rove and hister beetles search many types of microhabitats for small arthropod prey, especially among decaying organic matter. Whirligig beetles attack terrestrial insects trapped on the surface of ponds. Checkered, bark-gnawing, and some click beetles prey on wood-boring beetles and their larvae. Soldier beetle larvae consume mostly the bodily fluids of invertebrates, while some adults eat aphids and other soft-bodied insects.

MILLIPEDE PREDATORS

Several beetles prey only on millipedes in spite of their ability to defend themselves with noxious compounds that include benzoquinones, hydrogen cyanide, and hydrochloride. Undeterred by its chemical arsenal, glowworm larvae will briefly run alongside a millipede before coiling themselves around the front of its body to deliver a lethal bite just behind and underneath the head. Using its sharp, sickle-shaped mandibles the larva injects its own concoction of paralyzing toxins and digestive enzymes that disables the millipede's defense system and pre-digests its tissues. Immobilized, the millipede quickly dies as its internal organs and tissues are liquified. The glowworm larva pushes its way inside the millipede's body to consume all but its exoskeleton and defensive glands.

← Beetles in several families, including Carabidae, Scarabaeidae, and Phengodidae, attack millipedes and have evolved various strategies to circumvent their prey's defenses.

Predatory ground beetles in the genus *Promecognathus* also prey on millipedes. Rather than engaging in behaviors to minimize contact with their prey's chemical countermeasures, these beetles have instead evolved a high tolerance to cyanide. Using its long, sharp mandibles the beetle viciously bites the millipede between its thickened dorsal plates until it dies of blood loss and exhaustion.

Unlike most dung beetles, *Deltochilum valgum* from the lowland rainforests of Peru is also a millipede predator. The scarab first grasps the millipede with its middle and hind legs, then wraps its long, curved hind legs tightly around the prey's body. When the millipede ceases to struggle, the beetle disarticulates the body using its uniquely chisel-like *clypeus* and saw-like forelegs. Often decapitated during the process, the dead millipede is then pinned by the beetle against its shelf-like surface pygidium with one hind leg as it walks forward on the remaining five legs in search of a sheltered place to feed.

PARASITES AND PARASITOIDS

Among Coleoptera, the best known parasites are mammal-nest beetles (Leiodidae). The adults and larvae of some species spend nearly their entire lives as external parasites feeding on skin exudates of beavers or rodents. Parasites that always kill their hosts are called parasitoids. The parasitoid larvae of blister beetles attack grasshopper egg masses buried in the soil. Others invade the subterranean nests of solitary bees and consume brood, pollen, and nectar.

INQUILINES

Beetles that live among social insects, especially termites and ants, are called inquilines. Such inquilines exploit the nests of their hosts for various resources, including food, shelter, and egg-laying sites, and, depending on the beetle species, are variously integrated into the colony's social structure. For example, some rove beetles (Staphylinidae) have evolved structures and behaviors to such an extent that their termite or ant hosts will accept them as one of their own.

WATER BEETLES

The smooth, streamlined, and rigid bodies of predaceous diving beetles (Dytiscidae) and water scavenger beetles (Hydrophilidae) are well adapted for inhabiting various freshwater habitats. Their middle and hind swimming legs, flattened and fringed with setae, are used like oars to propel these beetles through the water. Dytiscids swim by moving their legs together in unison, while hydrophilids use their legs alternately.

Water beetles regularly return to the surface to expel carbon dioxide and replenish their supply of oxygen. Hydrophilids break through the surface tension headfirst with their antennae to draw a layer of air over the underside of their abdomens. Dytiscids trap a bubble of air under their elytra by breaching the water surface with the tips of their abdomens. With their oxygen supply nearly exhausted, dytiscids may expose the bubble from the tip of their elytra to function briefly as a physical gill to replenish its oxygen from the surrounding water.

↓ The front tarsi of male *Eretes sticticus* (Dytiscidae) are expanded with pads underneath that help to grip the female's elytra while mating.

↓ The nanostructure of adhesive pads has inspired the development of medical smart devices that continually monitor the biochemistry of sweat.

→ Whirligig beetles (Gyrinidae) prey on insects trapped on the pond's surface. The adhesive pads on the front feet of male *Dytiscus marginalis* (Dytiscidae) help them to grip the female's grooved elytra during copulation. Their larvae, called water tigers, roam the bottom of the pond in search of prey.

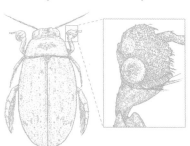

LEARNING THE LINGO

The terms introduced or adventive are applied to those beetles established outside of their native range by humans, either purposely (as in biocontrol agents) or by accident. Native species are indigenous to a particular area, while precinctive is applied to native or indigenous species that occur nowhere else. Coleopterists often use "endemic" to characterize native beetles with limited distributions, but some think that this term is best applied within the realm of epidemiology. Non-native populations do not occur naturally in a specified area. Invasive species are non-natives that were purposely or accidentally introduced by humans and are considered pests or potential pests.

Species richness simply refers to the number of beetle species that inhabit a defined ecosystem or region, while species diversity considers not only the number of species, but also the abundance of each species. Pest outbreaks occur when populations rise above normal levels and cause economic damage or threaten other human interests (see pages 102–103).

↓ The Japanese beetle, *Popillia japonica* (Scarabaeidae), native to East Asia, has invaded eastern North America and parts of Europe.

↓ The emerald ash borer, *Agrilus planipennis* (Buprestidae), has decimated hundreds of millions of ash trees in eastern North America.

→ Asian longhorn beetles, *Anoplophora glabripennis* (Cerambycidae), are native to eastern China and the Korean Peninsula. Likely introduced into North America as larvae in untreated wood packaging, scattered populations are also established across Europe. This destructive wood-borer kills maples as well as other hardwood trees.

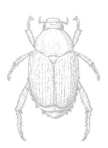

BEETLE ZOOGEOGRAPHY

A hallmark of beetles is their natural pattern of geographical distribution. Some occur only in tropical bioregions, such as equatorial rainforests and savannas. Others are known only from coniferous or mixed forests, grasslands, deserts and other temperate bioregions.

BIRTH OF BIOGEOGRAPHY

The distributions of animals, especially those of vertebrates, have been used to divide up the Earth into biogeographic realms characterized by distinctive floras and faunas. British zoologist Philip Sclater (1829–1913) established six zoological realms in 1858 that partly coincide with the continents. His Palaearctic, Afrotropical, Indian, Australasian, Nearctic, and Neotropical realms were based on the distribution of birds. British naturalist Alfred Russel Wallace, well known for his interest in beetles, revised this in 1876 based on the distributions of mammalian families worldwide. This work became a cornerstone across multiple biological disciplines, including beetle zoogeography.

GONDWANA

Isolated populations of related beetle taxa are considered to be relicts of formerly widespread populations. For example, widely disjunct populations of phylogenetically related scarabs that occur in the southern temperate regions of South Africa, Argentina and Chile, and Australia and New Zealand provide strong evidence that these southern continents once formed a supercontinent with Antarctica. Known as Gondwana, this ancient landmass began to break apart during the Jurassic about 180 Mya.

TERRESTRIAL REALMS

Most of North America, including the temperate regions of northern Mexico, comprise the Nearctic region. Most of Mexico, Central and South America, and the Caribbean make up the Neotropical realm. The Palaearctic Realm is comprised of North Africa, the temperate portions of the Arabian Peninsula, and Eurasia north of the Himalayas. All of sub-Saharan Africa, Madagascar, and most of the Arabian Peninsula are encompassed by the Afrotropical realm. Sometimes called the Oriental realm, the Indomalayan Realm stretches from India to southern China and most of Indonesia. The Australasian realm includes Australia, New Guinea, and the eastern portion of the Indonesian Archipelago.

CONTINENTS ADRIFT

Biogeographers have long noted the similarity of beetles in the Nearctic and Palaearctic realms, Madagascar and India, and southern South America with Australia and New Zealand. Plausible explanations for these disjunct distributions remained elusive until the early 20th century. German climatologist and geographer Alfred Wegener (1880–1930) noted that the Earth's continents fit together into continuous supercontinents like a jigsaw puzzle. In 1912, he proposed a hypothesis known as continental drift where continents slowly move across the Earth's surface, colliding and separating from one another over the course of geological time. Initially controversial, Wegener's concept gradually gained acceptance from the 1950s and became the foundation for the widely accepted scientific theory of plate tectonics.

IMPORTANT INSIGHTS

The biogeographical realms proposed by Sclater and Wallace continue to undergo revision based largely on the distributions and phylogenetic relationships of vertebrates. Zoogeographic research on beetles focuses mainly on family level data that incorporates not only distributional information, but also molecular phylogenetics and plate tectonics. Such analyses help to reveal the evolutionary origins of beetles, offer explanations for their current patterns of distribution, confirm past geological events in light of plate tectonics, and provide important insights into how beetles might respond to climate change.

CLIMATE AND OTHER LIMITATIONS

Where a particular beetle lives is determined by its habitat requirements. Habitats are shaped by ecological elements and historical influences. Ecological elements involve interactions between abiotic factors (for example, climate and soil type) and biotic factors (including food availability, competitors, and natural enemies). Historical influences include recent land use practices as well as geological phenomena such as glaciation and plate tectonics. Ecology determines if a beetle can survive in a given habitat, while the area's history determines its presence there in the first place.

Climate strongly influences beetle species richness and diversity. Warm tropical and subtropical regions support the greatest numbers of species, while diversity diminishes in cooler temperate and polar regions. Once predictable, seasonal fluctuations in these regions have given way to unseasonable heatwaves, cold snaps, flooding, and droughts. The varying abilities of beetles to adapt to the interactions of climate change, habitat destruction, and invasive species will determine which species survive and where.

↓ Outbreaks of the European spruce beetle, *Ips typographus* (Curculionidae), have impacted Norway spruce forests across southern and central Europe (see pages 104–105). Bark beetle outbreaks are altering the world's temperate coniferous forests.

→ The bronze carabid, *Carabus nemoralis* (Carabidae), is one of eight larger-bodied beetle species sampled in British Columbia that were found to have decreased in size in response to increasing autumnal temperatures over a 30- or 100-year time span.

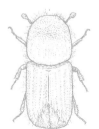

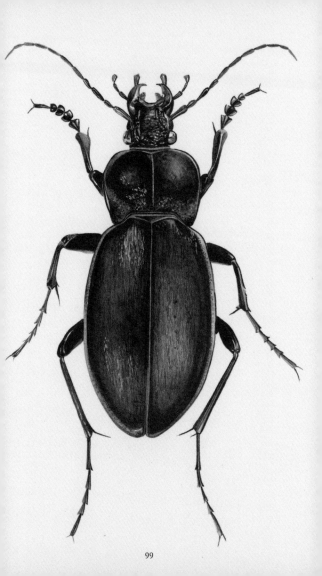

DISPERSAL

Beetles usually disperse over short distances by walking, running, or swimming. However, *Onymacris plana* from Namibia can cover 12 miles (19 km) in just ten days! Strong swimmers, predaceous diving beetles (Dytiscidae) are capable of colonizing new and sometimes distant water bodies thanks to their ability to fly. The African dung beetle *Pachylomera femoralis* flies long distances in search of fresh feces. Once airborne, smaller beetles are sometimes transported far away by the wind.

GROUNDED

Flightless beetles, such as those adapted for spending most of their lives deep in the soil, inside caves, or swimming in subterranean springs, all typically have limited powers of dispersal. More than 30 species of rain beetles (*Pleocoma*) are scattered along the Pacific Coast of North America. Although the males are strong fliers, the flightless females limit the distribution of each species.

↓ The large dung beetle *Pachylomera femoralis* (Scarabaeidae) occurs in deep sand habitats across southern parts of Africa.

MARITIME BEETLES

The adults and larvae of longhorn beetles (Cerambycidae) and weevils (Curculionidae) are sometimes capable of surviving long periods at sea inside drifting logs. Remote oceanic islands that typically lack insect diversity are sometimes disproportionately colonized by these and other wood-boring beetles.

HITCHHIKERS

A few beetles rely on other animals for transport, a strategy known as phoresy. By latching on to a solitary bee, the parasitic larvae of blister beetles gain access to the bee's nest and brood. Adult *Antherophagus* hitch rides on bumblebees in order to scavenge debris in their nests.

INTRODUCTIONS

Some beetles and their larvae are dispersed by human activity and are accidentally transported within nursery stock, packaged foods, and timber products. Beetles used as biological controls, especially in the families Scarabaeidae, Coccinellidae, Chrysomelidae, and Curculionidae, are deliberately introduced into new environments.

INVASIVE SPECIES

The movement of people and products has resulted in the purposeful and accidental introduction of plants, animals, fungi, and microbes worldwide. Some of these species eventually prove to be troublesome economically and/or environmentally. Unfettered by natural checks and balances, invasive species are free to dominate new habitats. Invasive insects, including beetles, pose serious challenges to sustainable economies and biodiversity by damaging food supplies, hindering trade, and endangering native species by disrupting natural ecosystems. *Parthenogenetic* pest weevils are particularly troublesome because only a single female is required to establish an infestatation.

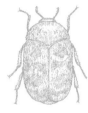

↑ Khapra beetles (Dermestidae) are small, secretive, and destructive.

KHAPRA BEETLE

Trogoderma granarium is one of the world's most destructive pests of stored grains and seeds. Native to Southeast Asia, this destructive beetle is now established throughout the Old World. Khapra beetle infestations contaminate and destroy human food supplies and disrupt trade as a result of quarantines restricting grain imports from countries known to have infestations.

IDENTIFICATION

Accurate and rapid identification is key to controlling invasive beetles. Genomic techniques can help identify the geographic origin of an invasive species, which can lead to the discovery and development of its natural predators, parasitoids, and pathogens for use as biocontrol agents to help manage beetle pests.

JAPANESE BEETLE

Popillia japonica is a serious horticultural and agricultural pest in parts of North America and Europe. The adults consume the flowers, leaves, and other vegetative structures of more than 300 species of ornamental and landscape plants, garden crops, and commercially grown fruits and vegetables. The grubs consume the roots of turfgrass and other plants, often causing severe damage. A warming climate will likely encourage the spread of this pestiferous beetle throughout the temperate regions of the world.

EMERALD ASH BORER

Agrilus planipennis is native to Northeast Asia, where it seldom causes any significant tree damage. First discovered in North America during the summer of 2002, emerald ash borers are now established throughout much of the Northeast and Upper Midwest where they have killed millions of ash trees. These tree killers have put nearly 100 species of insect herbivores dependent on ash trees at risk of high endangerment, including four horned scarab beetles: *Dynastes grantii*, *D. tityus*, *Xyloryctes jamaicensis*, and *X. thestalus*.

ASIAN LONGHORN BEETLE

Native to China and Korea, the Asian longhorn beetle, *Anoplophora glabripennis*, occurs in several regions in eastern North America. The tunneling activities of the larvae threaten millions of street and park trees and pose serious problems for the maple syrup industry in the Northeast. Eradication efforts require cutting down, chipping, and burning thousands of infested trees. This destructive beetle is also now established in at least 11 European countries. In recent years, the Asian longhorn beetle has increased its range in China due to the widespread planting of poplar hybrids that are susceptible to attack.

PALM PESTS

A serious pest of ornamental and date palms, the red palm weevil, *Rhynchophorus ferrugineus*, is the world's most destructive palm pest. A native of Southeast Asia, this very large weevil has become widespread in regions where palms grow. An infestation of a related species, *R. vulneratus*, discovered in southern California in 2010 was eradicated in 2015.

OUTBREAKS

Although difficult to assess, studies suggest that bark beetle outbreaks in western North America have occurred regularly for the past several centuries. Such outbreaks are important natural forces in forest ecosystems and play an essential role in their growth and regeneration. Beetles typically focused their attentions on older and less healthy trees, thus allowing younger, more healthy individuals to compete and thrive.

WHAT CAUSES OUTBREAKS?

Bark beetle outbreaks are the result of complex interactions between weather, climate, food availability, natural enemies, and other biotic factors. In Europe and North America, climate change and forest management practices have contributed to the increasing frequency and intensity of naturally occurring outbreaks of native species of bark beetles.

BARK BEETLES

Recent outbreaks of the Eurasian spruce bark beetle, *Ips typographus*, have decimated millions of coniferous trees in Europe. Only 4.0–5.5 mm in length, these small beetles quickly kill mature trees by overcoming their defenses with the aid of fungal symbionts that support the development of the beetle's larvae. Bark beetles apparently locate potential host trees by tracking odors produced by these fungal symbionts as they metabolize resin. These odors not only help bark beetles to locate potential feeding and breeding sites, but they may also help them to assess a tree's defense capabilities and determine the number of conspecific bark beetles already present on the tree.

Unlike previously recorded outbreaks in western North America, bark beetle infestations today are occurring simultaneously across numerous forest ecosystems, killing larger numbers of trees faster and over longer periods of time. Since 1990 bark beetles have destroyed forests by killing billions of trees in more than 150 million acres (61 million hectares) of forests managed for timber products, from Alaska to northern Mexico. Current elevated temperatures and drought levels, combined with vast expanses of mature trees in dense concentrations across the region, have contributed to the proliferation and expansion of bark beetles, especially a handful of tree killers in the genera *Dendroctonus*, *Ips*, and *Scolytus*. Warmer temperatures accelerate bark beetle life cycles and reduce their mortality rates during the winter. Prolonged warm temperatures and drought stress can weaken trees, curtailing their ability to defend against bark beetle attacks.

↑ Warmer summer and winter temperatures due to climate change drive bark beetle outbreaks that have a serious impact on carbon cycles and forest ecosystems.

ECOLOGICAL REQUIREMENTS

In order to survive, each beetle species occupies one or more habitats characterized by a particular suite of vegetation, soil type, and climate that meets the requirements of all four of its developmental stages in terms of food, water, shelter, and egg-laying sites.

Widespread species, such as *Phanaeus vindex,* are tolerant of various environmental conditions and capable of utilizing many kinds of foods. This dung beetle inhabits diverse wooded and grassland habitats with clay or sandy soils, from the eastern United States westward to southeastern Arizona and northern Mexico. Both adults and larvae can utilize many kinds of vertebrate feces, including that of humans.

Beetles tolerant only of narrow environmental conditions and restricted to a single mountain range, dune complex, or cave system are often in need of conservation. For example, *Colophon westwoodi*, a flightless stag beetle that occurs only on Table Mountain in South Africa, is listed as vulnerable by the International Union for Conservation of Nature (IUCN).

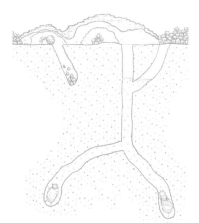

↙ Both sexes of *Phanaeus vindex* excavate tunnels beneath a dung pat and construct one or more balls made of moist dung coated with soil. Each brood ball serves as food for the larva developing within, while an additional food ball is provided for freshly emerged adults.

→ Once known colloquially as the royal tumbleturd in Colonial America, males of *Phanaeus vindex* possess an impressive horn that sweeps back over their broad thoracic shield.

FORESTS

The woody tissues of forest trees and shrubs are utilized as food by beetles. Injured and dying trees support an incredible diversity of species that often have specific food preferences for living, dead, or decaying bark and heartwood. Leaves, stems, flowers, cones, seeds, and roots also provide food and shelter for many other species.

HERBIVORES AND FUNGIVORES

Nocturnal scarabs munch on the edges of leaves and may be so abundant that they can occasionally cause severe defoliation. Although some beetles are generalists and feed on a diversity of plant species, others are specialists that will only eat tissues from trees in the same genus or family. In Australia, Christmas beetles in the genus *Anoplognathus* are sometimes conspicuous defoliators of eucalyptus trees, although their numbers around densely populated areas have dwindled considerably over the past few decades. Chrysomelid beetles regularly feed on leaves as both larvae and adults. Weevils and their relatives may also be abundant feeding on various vegetative tissues, both day and night. Although the larvae of a few species feed externally on leaves, most develop inside seeds and stems, or in the soil among

---

PREDATORS

Lady beetles are commonly found in the canopy preying on sap-feeding aphids and phytophagous mites. Predatory ground beetles and click beetles scour bark surfaces for prey, day and night. In the tropical rainforests of Australia and Southeast Asia, tree trunks encrusted with lichen and moss serve as hunting grounds for Tricondyla and other arboreal tiger beetle genera. Adults of some checkered beetles are typically found running along trunks and branches infested with bark beetles and other wood-boring beetle prey, while others are regularly encountered on flowers.

roots. Shiny flower beetles, sap beetles, silky fungus beetles and other cucujoids are common on leaf surfaces, where they likely feed on various fungi invading the tissues of dying leaves. At night, cryptic species of fungus weevils graze on patches of fungi growing on bark.

FLOWERS

Beetles in several families are regularly found feeding and mating on flowers. Of these, jewel beetles and flower chafers are among the most conspicuous and beautiful species, especially in the Southern Hemisphere. Flower-visiting jewel, net-winged, soldier, and longhorn beetles are often defensively bee- or wasp-like in appearance and manner.

SAPROXYLIC SPECIES

Although some darkling beetles and their relatives are occasionally found up in the canopy as adults, the adults and larvae of many forest-dwelling species make their living in the decaying wood of snags (standing dead trees) and stumps, or among the chunks of woody debris and leaf litter that accumulate at their bases. These *saproxylic* beetles make up complex assemblages of mostly secretive species representing dozens of families that depend on the wood of dead and dying trees and their associated fungi. Saproxylic beetles represent a significant part of forest biodiversity and play a key role in nutrient recycling and environmental health. They are considered important indicator species, especially in temperate old-growth forests in Europe and North America. Forestry practices that cull trees before they mature diminish the accumulation of woody debris properly seasoned with fungi, thus negatively impacting its associated saproxylic beetle fauna.

→ Dead logs are alive with fungi as well as saproxylic beetles and other insects that depend on decaying wood and associated microorganisms in order to survive.

WETLANDS AND SHORELINES

Beetles inhabit an incredibly diverse array of aquatic and semiaquatic environments, including cold mountain streams, ponds and lakes, brackish marshes and estuaries, shorelines, and intertidal coastal habitats.

SWIMMERS AND CRAWLERS

While some aquatic beetles favor cold, fast streams, others prefer ponds or pools along the edges of slow-moving streams. Based on their modes of locomotion, aquatic beetles are divided into two basic groups: swimmers and crawlers. Swimmers move through the water column and along vegetated or graveled substrates using their powerful natatorial legs. The best swimmers, such as species of *Dytiscus* and *Hydrophilus*, propel their smooth, streamlined bodies with their oarlike middle and hind legs fringed with setae. However, equally agile whirligig beetles spend their lives singly or in groups on the surfaces of ponds or along the relatively still edges of streams and rivers.

LIFE ON THE EDGE

The shallows along the edges of various bodies of standing and flowing freshwater habitats that support emergent plants comprise the littoral zone. This zone is often inhabited by less able swimmers, such as crawling water beetles, as well as beetles lacking natatorial appendages altogether. Riffle and long-toed water beetles are typically found in streams, creeping on algal mats or clinging to rocks and logs.

The shorelines of ponds, rivers, and streams form a natural transition between aquatic and terrestrial habitats. Beetles living on the edge of these habitats often represent terrestrial species with aquatic tendencies, such as *Omophron* (Carabidae), or vice versa, as evidenced by *Hydraena*, *Ochthebius*, and other minute moss beetles (Hydraenidae).

WHIRLIGIG BEETLES

Whirligig beetles deftly avoid predators partly because of their ability to see above and below the water's surface and as a result of powerful natatorial middle and hind legs that enable them to swim rapidly at speeds of 21–57 in (53–144 cm) per second. The tip of their abdomen functions as a rudder and can bend downward nearly 90 degrees. They also have pygidial glands that produce an oily and noxious fluid consisting mostly of sesquiterpenes that makes them distasteful to fish. Their exoskeleton continuously secretes a waxy outer layer that repels water molecules, giving them a little extra zip as they swim across the water.

↑ Whirligig beetles (Gyrinidae) swim on the surface of still waters, but will dive beneath the surface for short periods when threatened.

DESERTS

The world's major non-polar deserts cover about one-third of the Earth's land surface. Characterized by sparsely vegetated mountains, dune fields, dry riverbeds, and alkali flats, these arid habitats support a surprisingly rich beetle fauna. To survive such harsh environments, desert beetles employ various morphological and behavioral adaptations that help keep them cool and hydrated.

FLIGHTLESSNESS

Darkling beetles are conspicuous insects that are typically black and heavily armored and possess several morphological and behavioral adaptations that enable them to survive in desert habitats. Foremost among their morphological adaptations is having lost the ability to fly. Thick and fused together, elytra fit snugly over the beetle's body to form the subelytral cavity. Water vapor released through the spiracles during respiration remains within the subelytral cavity, thus preventing desiccation. The reduction or loss of flight wings also increases the capacity of the subelytral cavity to accommodate the ingestion of more food and egg development, qualities that enhance the beetles' survival. Some desert-dwelling dung scarabs and hide beetles, as well as many weevils, are also flightless.

THERMOREGULATION

Desert tenebrionids sometimes beat the heat by burying themselves in sand, hiding under rocks, or resting in mammal burrows where temperatures are significantly lower and relative humidity higher than on the surface. Other species avoid extreme seasonal temperatures altogether by restricting their surface activities to the winter. Clothed dorsally in long setae with waxy filaments caked in sand, some species are able to block ultraviolet radiation and seal in moisture. Still others rely on a waxy layer covering their bodies or white elytra created by air spaces within the cuticular layers to help them regulate their body temperature. For example, the body of the desert ironclad beetle, *Asbolus verrucosus*, which inhabits the Mojave and Sonoran Deserts of the American Southwest, is coated with bluish white, waxy filaments that help keep the beetle cool and prevent the loss of precious moisture through their exoskeleton. Several leggy diurnal species of *Onymacris* and *Stenocara* in the Namib desert have elytra that are partially or completely white as a result of air spaces within the layers of cuticle. Adaptations that lighten the color of the exoskeleton were once thought only to facilitate thermoregulation, but recent research suggests that these paler hues may serve cryptic or aposematic functions.

WATER ACQUISITION

Several beetles that inhabit the Namib desert have evolved interesting behaviors that enable them to acquire water from night-time coastal fog. Species of *Lepidochora* use their flat, saucer-shaped bodies to bulldoze long trenches in the sand that are parallel to the prevailing wind. These thirsty beetles later return to drink water captured by the ridges flanking the trenches.

→ The button beetle, *Lepidochora discoidalis* (Tenebrionidae), is one of three species in the genus that inhabit the coastal sand dunes of the Namib desert.

EXTREME HABITATS

T hermal springs, saline waters, subterranean aquifers, and caves also harbor beetles. Species encountered in thermal springs and saline waters are often tolerant of a wide range of environmental conditions. However, *Hydroscapha redfordi*, named after the well-known actor and conservationist, is known only from algal mats covered by thin layers of water flowing from two hot springs in northeastern Idaho. A predaceous diving beetle, *Hygrotus salinarius*, is restricted to saline pools of the northern Great Plains of North America.

LIFE UNDERGROUND

Subterranean predaceous diving beetles are found worldwide. Generally pale and flightless, these beetles typically have eyes and swimming setae that are both reduced or entirely absent. In Texas, the Edwards Aquifer diving beetle, *Haideporus texanus*, is one of five subterranean diving beetles known only from water-filled spaces in karst limestone of the Edwards (Balcones Fault Zone) Aquifer.

CAVE DWELLERS

Troglobitic beetles are known mostly from regions that never experienced glaciation, live at low population densities, have relatively long lives, and produce very few offspring. Typically pale and sometimes blind, they often have elongated antennae and long sensory setae called trichobothria that sense changes in air currents, indicating danger or the presence of prey. Very long, slender legs enable troglobitic species to forage or hunt over long distances. Although some species rely on bat droppings or the food supplies of other vertebrates living near cave entrances, species living deep inside caves are totally dependent upon the remains of other cave organisms or organic materials washed in by streams and seeps.

ENDANGERED SPECIES
OF THE EDWARDS AQUIFER

Texas has nine major aquifers that provide diverse habitats for rare vertebrates and invertebrates. One of these groundwater systems, the Edwards Aquifer in south central Texas, is home to several beetles associated with its subterranean springs and karst habitats that are listed as endangered by the U.S. Fish and Wildlife Service. These species include the Comal Springs riffle beetle (*Heterelmis comalensis*), Comal Springs drypoid beetle (*Stygoparnus comalensis*), Helotes mold beetle (*Batrisodes venyivi*), and two ground beetles, *Rhadine exilis* and *R. infernalis*. Changes in water quality and quantity can negatively impact these and other animal populations adapted to these usually stable subterranean environments.

↑ Eyeless and depigmented, the subterranean Edwards Aquifer diving beetle, *Haideoporus texanus* (Dytiscidae), occurs in the springs of the Edwards-Trinity Aquifer of Texas.

SYNANTHROPIC BEETLES

Synanthropic beetles benefit from their association with us by taking advantage of various resources with which we permanently surround ourselves. Some have become nearly cosmopolitan pests in our homes, museums, warehouses, and croplands. Already adapted to feeding on similar foods in natural environments, these beetles readily exploit our stored foods, crops, and other organic materials.

Flightless Ptinidae that scavenge organic materials in various animal nests regularly infest human food stores. Wood-boring ptinids that develop in dead tree limbs easily take advantage of furniture, flooring, and other wood products in our homes. Insect remains trapped in spider webs or pinned and labeled in museum collections are all the same to hungry *Anthrenus* larvae. Larval ham beetles in the genus *Necrobia* are predators that bore into dead flesh and see little distinction between fly-infested roadkill and provisions of dried meat. Some leaf beetles and weevils have expanded their taste for native plants to include closely related species in cultivation.

↓ Drugstore beetles, *Stegobium paniceum* (left), and cigarette beetles, *Lasioderma serricorne* (right), are both in the family

Ptinidae. They eat all kinds of stored plant and animal materials and are common pests in food processing plants and warehouses.

→ The varied carpet beetle, *Anthrenus verbasci* (Dermestidae), was originally described by Carl Linnaeus in 1767. Although their larvae utilize animal protein as food, the adults prefer pollen and nectar, and are shown here feeding on the flowers of ground elder (*Aegopodium podagraria*).

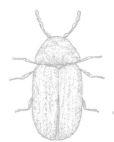

EDIBLE BEETLES

Entomophagy, the consumption of beetles and other insects, is widely practiced outside North America and Europe. Interest in sustainable farming practices, coupled with the relatively low cost of raising insect-based protein compared with traditional sources of animal protein, has substantially increased the appeal of producing insects as food.

Yellow mealworms, palm weevil grubs, and various other beetle larvae are among the more than 300 species of beetles eaten around the world. They are baked, fried, roasted, or toasted, then seasoned with various spices. To make the idea of eating beetles more appealing to markets in North America and Europe, companies are using dried mealworms ground into a protein-rich flour as a primary ingredient in otherwise traditional baked goods. Compared with conventional protein sources (beef and other meats), mealworm flour contains more than twice the amount of protein but requires only a fraction of the water to produce and contributes virtually no greenhouse gases.

↙ Red palm weevil grubs, *Rhychophorus ferrugineus* (Curculionidae), are eaten alive, toasted, or roasted on skewers in Vietnam.

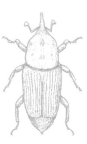

→ In New Zealand, the larvae of the huhu beetle, *Prionoplus reticularis* (Cerambycidae), are high in nutrients. With a peanut butter-like consistency and buttery chicken flavor, they are considered a traditional delicacy by the Maori people and a food challenge for tourists.

MYTHS AND FOLKLORE

Lacking the ability to clearly see and observe the activities of beetles, ancient peoples sometimes viewed conspicuous species as magical, imbued them with mythological powers, and used them as symbols of good and evil.

SCARAB BEETLES

The best known mythological beetle is the sacred dung scarab, *Scarabaeus sacer*. To the ancient Egyptians, its curious dung-rolling behavior symbolized the morning sun god Khepri, who moved the sun like a dung ball across the heavens. Scholars have suggested that the ancient Egyptians' rudimentary understanding of the larva's transformation into a mummy-like pupa within a buried dung ball may have inspired the construction of the pyramids and the many chambers within to prepare their dead for the afterlife. Scarab images appeared everywhere, from hieroglyphs to funerary carvings and amulets. Interestingly, medieval Christianity viewed dung scarabs and their association with waste as symbols of bad luck, foulness, sin, and evil.

MYTHOLOGICAL LINKS

Mythology permeates coleopteran taxonomy. *Dynastes hercules* was named for the Roman hero known for his great strength, while *D. neptunus* was inspired by the god of the sea. *Goliathus atlas* is derived after both a biblical character and the Titan condemned to bear the weight of the heavens on his shoulders. *Sisyphus* is a genus of dung-rolling beetles named after the king sentenced by Hades to forever roll a boulder uphill that always rolls back down before reaching the summit. The Roman goddess of beauty and love inspired the name *Termitotrox venus,* while her son with Mars suggested the specific epithet in *T. cupido.*

THE MIDDLE AGES

During the Middle Ages, select species of Coccinellidae were dedicated to the Virgin Mary and named "beetles of Our Lady." With the passage of time, these beetles became known in the English-speaking world as ladybird beetles, lady beetles, and ladybugs. The popularity of the brightly colored beetles remains strong to this day, and they are regularly featured in children's books and nursery rhymes.

THE MAYA

Several other kinds of beetles have served as symbols, especially fireflies. The Maya associated fireflies with the stars, lit cigars, and gods. Their flashing lights are also considered to be signs of good luck or variously associated with spirits of the dead.

↑ Seven-spotted lady beetles, *Coccinella septempunctata* (Coccinellidae), and other coccinellids often feature in folklore and children's nursery rhymes.

BEETLES THAT INSPIRE

Biomimetics, or biomimicry, is the study of nature and natural processes. The goal of biomimetic studies is to gain an understanding of natural mechanisms, then apply that knowledge to solve human challenges. Rather than developing complicated and expensive engineering techniques from scratch that will likely require extended periods of trial and error, bioengineers incorporate into their designs biological processes forged by millions of years of natural selection.

Given their persistence as a group for nearly 300 million years, it should come as no surprise that beetles embody a wealth of scientific and technological inspiration. Each species contains morphological, genetic, and chemical characteristics that inspire the development of novel materials, compounds, and other disruptive technologies to advance the fields of medicine and technology.

FIREFLIES AND CLICK BEETLES

Detailed studies on the light-producing organs of fireflies and click beetles have not only led to breakthroughs in the development of brighter and more energy-efficient methods of lighting, but also to the development of tests for detecting bacterial contamination in food and drink, studying gene expression and cell physiology, imaging tissues for biomedical research, and detecting organic compounds associated with possible life in outer space.

ADHESIVE POWER

The ability of beetles and other insects to walk on walls and ceilings has long fascinated evolutionary biologists and bioengineers. A team of German and American scientists examined more than 300 species of insects in search of those that possessed the ideal properties for a new kind of adhesive tape that was glue-free. They initially focused on two independently evolved types of sticky insect feet—the smooth pads of grasshoppers, and beetles with densely "hairy" feet. Using beetle feet as a model, the researchers developed a glue-free tape that adhered to surfaces by means of high-density, mushroom-shaped nanostructures that could easily adapt to uneven surfaces and other irregularities, just like hairy beetle feet.

MECHANICAL STRENGTH

Joining dissimilar materials with adhesives, mechanical fasteners, or welding has long been a challenge for engineers because these technologies can add weight or introduce stress that leads to corrosion and fractures. Looking to nature for solutions, bioengineers studied the diabolical ironclad beetle, *Phloeodes diabolicus*, a flightless beetle with an incredibly hard body. Microscopic images, computer simulations, and 3D-printed models reveal that the strength of their heavily armored exoskeleton is the result of interlocking plates reinforced with impact-absorbing structures capable of withstanding

↑ The zipper-like locking mechanism binding the elytra of diabolical ironclad beetles, *Phloeodes diabolicus* (Zopheridae).

up to 39,000 times the beetle's own body weight. For example, the exoskeleton's dorsal surface is reinforced by a series of tough interdigitated structures consisting of interconnected zipperlike teeth and damage-resistant, jigsaw-like protrusions. Each of these structures is composed of layers of tissue glued together by proteins. Bioengineers are exploring the possibility of synthesizing comparable structures to be used as fasteners in the structural joints in buildings and bridges, as well as the manufacture of aircraft engines.

ONGOING RESEARCH

Beetles embody an exciting frontier for discovery by bioprospectors, bioengineers, and biomaterials scientists, a frontier that we have only just begun to explore. The discovery and subsequent synthesis of beetle-inspired compounds, structures, and materials not only requires significant simultaneous investment in sophisticated analytical tools and technologies, but also in taxonomic and systematic research and conservation of beetle habitats.

ARTS AND CRAFTS

The forms and colors of beetles have inspired artisans and craftspeople alike for centuries. Scarabs commonly appeared in the religious art of ancient Egypt. The borders of medieval manuscripts were often illuminated with beetles. During the Renaissance, beetles appeared in still-life paintings and various decorative arts, but German artist Albrecht Dürer's unprecedented *Stag Beetle* (1505) turned them into a focal point, a trend that morphed into scientific illustration during the Age of Enlightenment.

BEETLEWING ART

Beetlewing art is an ancient craft technique that incorporates permanently iridescent elytra. While mourning their dead, people in Myanmar and northern Thailand don shawls fringed with the brilliant green elytra of *Sternocera* jewel beetles. The Victorian actress Dame Ellen Terry famously wore a similarly adorned beetle dress while playing Lady Macbeth and posed in this costume for American painter John Singer Sargent. Earlier this century, Belgian artist Jan Fabre affixed the elytra of nearly half a million jewel beetles to the ceiling of the Royal Palace in Brussels.

JEWELRY

The durable horns, mandibles, elytra, and legs of large, dead beetles are often fashioned into jewelry. People in various parts of the world adorn themselves with living bejeweled or bioluminescent species.

MODERN ART

The unsanctioned works of English street artist Banksy sometimes straddle the interface between art and vandalism. His *Withus Oragainstus, United States*, an anti-war piece featuring a harlequin beetle festooned with military hardware and mounted in a shadow box, was discovered in the Hall of Biodiversity in New York's American Museum of Natural History.

ZOPHERUS CHILENSIS

The ma'kech, *Zopherus chilensis*, is a slow-moving and long-lived saproxylic beetle that occurs from southern Mexico to Columbia and Venezuela. In the Yucatán Peninsula gemstones are affixed to their very hard backs, before they are tethered to clothing with a short chain and pin as a good luck charm. Sold to tourists as a novelty, the ma'kech is a reminder of an ancient Yucatecan legend where a young Mayan prince was transformed into a beetle to elude the guards of his lover. Efforts are underway to develop a captive breeding program that will help conserve wild populations of this beetle.

↓ More than a million iridescent green elytra of *Sternocera aequisignata* (Buprestidae) from Thailand decorate the Royal Palace in Brussels.

BEETLES AND POPULAR CULTURE

To call attention to the diversity of beetles and the importance of their identification and classification, some coleopterists have drawn on various elements of popular culture for inspiration in naming new species. Science fiction, comic book, and Pokémon characters, as well as the names of athletes, movie stars, and politicians, have all inspired the scientific names of beetles.

In 2018, three scarab species, *Gymnetis drogoni*, *G. rhaegali*, and *G. viserioni*, were named after dragons appearing in the *Game of Thrones* series. The following year in a study of *Trigonopterus* weevils from Sulawesi, the authors, inspired by pop culture sources that included *Asterix* comics and the Star Wars franchise, named their new weevil species *T. astrix*, *T. idefix*, *T. obelix*, and *T. yoda*; another species, *T. chewbacca* from New Britain (an island in the Bismarck Archipelago in the western Pacific Ocean), was described in a separate study. More recently (2020), three species of Australian fire-colored beetles were named after three rare Pokémon birds: *Binburrum articuno*, *B. moltres*, *B. zapdos*.

↓ The epithet of flightless weevil *Trigonopterus chewbacca* (Curculionidae) comes from the dense scales on its head and legs.

↓ *Demyrsus digmon* (Curculionidae) bores into cycad trunks in Queensland, Australia, its name inspired by a Japanese anime figure.

→ The habits of *Gymnetis viserioni* (Scarabaeidae), known from Panama, Columbia, and Ecuador, are unknown, but they likely feed on decaying fruits and sapping tree wounds. Their variable color pattern is disruptive and probably renders them less beetle-like to hungry predators.

BEETLES AS MEDICINE

The use of insects and insect-derived products as resources for medicine, or entomotherapy, has long been practiced by humans worldwide. One of the earliest known uses of beetles in medicine occurred in the Mediterranean region where the large mandibles of living *Scarites* ground beetles were applied to close wounds until they healed. Other beetles have been used live, cooked, ground, or prepared as infusions, plasters or bandages, salves, and ointments to treat or prevent a broad range of ailments.

CURES AND TREATMENTS

Pulverized ladybirds, also known as lady beetles or ladybugs, were once sold in Europe to relieve toothache and as a cure for measles and colic. Oil extracted from the larvae of the cockchafer *Melolontha vulgaris* was applied topically to treat scratches and as a cure for rheumatism, while the wine-soaked adults were ingested as a treatment for anemia. Whether the benefits were real or imagined, these and other uses of beetles as medicine likely originated with them as food sources.

The large mandibles or horns of some beetles have long suggested their use as libido boosters. Eating the ashes of male European stag beetles (*Lucanus cervus*), was once thought to be an effective sexual stimulant. More recently, men of the Hñ ähñ tribe living in the state of Hidalgo, Mexico, enhance their virility by eating male ox beetles, *Strategus aloeus*, especially their bladelike thoracic horns. Any positive effects experienced by those consuming these beetles are simply due to the power of suggestion.

Today, the use of beetles and other arthropods in traditional medicine in East Asia is commonplace. Traditional medicine in China utilizes nearly 50 species of beetles, including blister beetles in the family Meloidae. Blister beetles contain the extremely toxic chemical compound cantharidin that is best known for its purported qualities as an aphrodisiac and for inducing abortions. Extremely toxic at low doses, cantharidin causes inflammation of the gastrointestinal tract and may cause death as a result of kidney failure. In spite of its toxicity, cantharidin is used widely in East Asia to treat skin boils, fungal infections, paralysis caused by strokes, swollen lymph nodes, rabies, gonorrhea, and syphilis. Extracts from *Hycleus* blister beetles are used in China as remedies for infectious fevers, scrofula, necrotic tissue, bladder stones, baldness, bruises, and urinary blockage. Rural people in South America and also dermatologists in the West have long used cantharidin topically to treat warts (human papillomavirus, or HPV) and water warts (*molluscum contagiosum* virus). Recent studies demonstrate that cantharidin and its derivatives inhibit several kinds of human cancer cells in the laboratory.

↑ Cantharidin from the blister beetle *Hycleus phaleratus* (Meloidae) is used in traditional Chinese medicine to treat a variety of ailments.

Another incredibly potent beetle toxin is pederin, which is found primarily in the hemolymph of female *Paederus* rove beetles. A by-product of an endosymbiotic *Pseudomonas* bacterium, pederin shows potential for use as an antineoplastic chemotherapeutic because it slows the growth of cancerous tumors.

FUTURE PHARMACOLOGY

In their own defense, beetles produce chemical compounds with pharmacological properties that include antibiotics, antifungals, antineoplastics, antimicrobials, anti-inflammatories, antioxidants, cytotoxins, and neurotoxins, but relatively few of these have been evaluated experimentally. Evidence that these substances might be useful for treating various bacterial infections, HIV and other viral diseases, and cancer bolsters the argument for the conservation of beetles and their habitats.

PHOTOGRAPHY AND
CITIZEN SCIENCE

With a bit of patience and practice, good macro photographs of medium- to large-sized beetles are relatively easy to make, thanks to smartphones with macro capabilities. Images of beetles against simple, uncluttered backgrounds look best, especially those that depict beetles engaged in natural behaviors such as feeding and mating.

Consider sharing your images on iNaturalist.com, where the community can assist with species identification. Such images may become part of one or more ongoing scientific projects documenting beetle distribution and diversity. Identifying the species in your photos and knowing that your images are contributing toward a greater understanding of the natural world will enhance your enjoyment of both beetles and photography.

Photographs alone may be insufficient for identification as many species require dissections of their reproductive organs and other difficult-to-see characteristics. Thus, you may find it desirable to collect beetles in order to carefully examine them through a hand lens or stereo dissecting microscope.

↓ Smartphones have become useful tools for taking images of beetles and identifying them in the field.

↓ A small hand lens or pocket loupe is an essential piece of equipment for examining beetles in the field.

→ The best images of beetles are made with digital DSLM or DSLR cameras equipped with a dedicated macro lens of a focal length of 50 mm or more. Macro photography requires patience and practice, but the results are definitely worth it.

COLLECTING BEETLES

The greatest challenge to studying trends in beetle diversity, abundance, range, occurrence, and other metrics is the lack of baseline data. Unlike butterflies, macro-moths, dragonflies, and damselflies, which are usually readily identified on sight, most beetles must be in hand to facilitate accurate species identifications. Therefore, collecting specimens is essential for scientific study and conservation.

Observing and collecting beetles provides valuable insights into their behavior, morphology, and classification, and helps to connect people with nature. Beetle collecting, especially in children, has sparked the careers of many entomologists, biologists, and other scientists, including the author of this book.

USEFUL EQUIPMENT

Collecting equipment and supplies can be made at home or purchased from entomological supply houses. Nets are useful for capturing beetles on the wing, sweeping them from vegetation, and scooping them from water. Beating sheets capture beetles dislodged from leaves and branches hit with a stick. Lights, especially UV lights suspended in front of an upright white bed sheet or wall, attract many beetles on warm spring and summer nights. Carrion, fruit, and feces lure select species, too. Vials and jars are essential for temporarily housing specimens. Forceps and aspirators enable easy handling of small specimens collected by these and other methods. A 10× pocket lens is helpful for examining specimens in the field and at home.

A pocket notebook is handy for keeping detailed field notes that include the date, time, temperature, locality, host plant, and other collecting information. These data, along with the name of the collector, will eventually go on the locality label affixed to each specimen. Museum and university websites offer detailed information on collecting beetles and other insects as well as how to properly kill, prepare, and store museum-quality specimens.

LOOKING FOR AND IDENTIFYING BEETLES

Look for beetles in backyards, vacant lots, and parks. Sample as many habitats as possible to increase the diversity of beetles in your collection. Forests, deserts, grasslands, savannas, coastlines, and wetlands all have a particular beetle fauna comprised of species with populations that wax and wane with the seasons.

There are websites, such as iNaturalist, where good images of beetles and other insects accompanied by locality information may be submitted for identification by knowledgeable members of the community. Several pages on Facebook are dedicated to beetles and their identification, too. Some museums post images online of their types, the actual specimens upon which coleopterists based their formal descriptions of beetles. Type specimens have a special function in zoological nomenclature and serve as the physical standard for species that are formally described in the scientific literature.

CLOSER EXAMINATION

Although some beetles are reliably identified by comparing specimens with photos, most require examination of their reproductive organs and other microscopic structures. Accurate identifications are best accomplished by examining specimens through a stereo dissecting microscope while consulting with pertinent scientific literature. Published monographs and other articles that offer illustrated identification keys are often accessible only through paid subscriptions or to persons with institutional library affiliations. However, some authors post their work at ResearchGate or you can email them directly and request a pdf of their work. Older, yet still important publications are often freely available online at the Biodiversity Heritage Library (BHL) and Journal Storage (JSTOR).

→ The identification of beetles often requires detailed microscopic examinations of both the external and internal structures of carefully prepared specimens.

BEETLE COLLECTIONS

More than a bunch of pinned dead insects, beetles deposited in publicly accessible museum and university collections provide researchers with voucher specimens that offer verifiable and permanent records of beetle diversity. Locality label data associated with each specimen contributes to our understanding of not only their distributions past and present, but also their habitat preferences and periods of activity.

Collections of beetles and other insects have long been used primarily for the purpose of identification, such as assisting entomologists with identifying invasive species that might threaten agricultural crops and forests managed for timber. These collections are vitally important resources for morphological investigations, especially in relation to studies on systematics and phylogenetic relationships.

SCIENTIFIC IMPORTANCE

Today, beetle collections have become incredibly useful tools for tracking changes in populations and habitats over time, thus increasing our understanding of their ecological and evolutionary responses to climate change. With the aid of advanced laboratory techniques, voucher specimens are also analyzed to identify pollutants and other chemical compounds on and in their bodies. Such analyses help to reveal environmental conditions, past and present.

Each beetle specimen offers a unique opportunity for researchers to investigate its morphological, biochemical, and genomic properties, all of which reveal its developmental and ecological history. These unique data sets embedded in meticulously prepared and curated beetle collections also link identity with a time and place, information that is vitally important to conservationists monitoring population trends, determining levels of endangerment, and developing hypotheses on potential impacts of climate change.

← Meticulously prepared specimens in museum collections are invaluable tools for identification, phylogenetic investigations, and conservation work.

ENTOMOTOURISM

A subset of ecotourism, entomotourism utilizes various kinds of educational programming to increase awareness of the ecological and socioeconomic roles of insects. Butterfly houses and insect zoos attract millions of entomotourists annually and provide insect encounters under controlled settings. People also travel to exotic locales to observe and appreciate insects in the wild, especially butterflies and dragonflies.

Sites with healthy populations of charismatic beetles, especially fireflies and dung beetles, have become popular tourist destinations, too. Firefly tourism has long been popular in Japan, Mexico, Taiwan, Thailand, and the United States, where synchronous flashing fireflies attract millions of people annually. South Africa's Addo Elephant National Park not only harbors one of the continent's southernmost populations of elephants, but it is also one of the last remaining habitats for the large and flightless dung beetle, *Circellium bacchus*. Visitors to the park have the opportunity to appreciate iconic African wildlife that includes the continent's largest dung-rolling beetle.

↓ Observing large, dung-rolling beetles (left) and flower-visiting monkey beetles (right), both in the family Scarabaeidae, are supplementing more traditional game-viewing and birding activities in South Africa, thus raising awareness of invertebrate conservation.

→ Firefly tourism is a rapidly growing industry in the United States and elsewhere in the world. The rush to experience the bioluminescent spectacle produced by these delicate beetles requires careful management to minimize harm to them and their environment.

BEETLES AND CLIMATE CHANGE

The impacts of human-caused climate change are upon us. Among insects, evidence of climate change is best supported by the shifting distributions of butterflies and dragonflies living in temperate habitats, especially those species living at the northern limits of their distributions.

DISPERSAL PATTERNS

Studies of the habits and distribution of European beetles, past and present, may provide useful insights into climate change and its impacts. For example, examination of Quaternary beetle fossils suggests that dispersal was the primary mechanism by which species adapted to episodes of changing climate in Europe about 10,000 years ago. As temperatures rapidly warmed at the end of the last glaciation, many species were forced to disperse northward or into suitable habitats at higher elevations. Although difficult to prove based on fossil evidence alone, species unable to migrate or adapt likely became extinct.

HABITAT FRAGMENTATION

Today, dispersal alone is unlikely to ensure the survival of many beetle populations because of habitat fragmentation due to development and agriculture. Habitat specialization and food preferences all play a critical role in a species' ability to adapt to short- and long-term changes. Drought and other conditions exacerbated by climate change can fundamentally alter habitats upon which beetles depend for food and egg-laying sites, especially in remaining habitat fragments. Over the course of evolution, fragmentation of habitat as a result of human activity is a very recent phenomenon and its long-term impacts on beetles will likely not become apparent anytime soon.

↑ Monitoring beetles adapted to the cold, such as *Diacheila arctica amoena* (left) and *Pterostichus macer* (right), both in Carabidae, can help us understand climate change impacts.

EFFECTS OF CLIMATE CHANGE

Scientists are studying the ways different animals and plants respond to climate change. Preliminary evidence indicates that beetles raised under controlled conditions decreased in size as temperature increased. Researchers at University of British Columbia analyzing the elytral lengths of Canadian beetle specimens revealed that select larger ground beetles have shrunk in size as much as 20 percent for nearly five decades as fall temperatures steadily increase. Whether the reduction in size results from climate change, as-of-yet imperceptible changes in food quality or quantity, increased predation, or other changing environmental factors likely varies among different beetles species and requires further study.

THREATS AND CONSERVATION

Beetles have a public relations problem. Even though they are the largest group of animals on Earth and thus represent a significant chunk of the world's biodiversity, many people still relate to them as being decidedly alien. Relatively small and sometimes pestiferous, beetles just don't elicit the same feelings of sympathy and moral responsibility that we afford larger and more charismatic animals such as pandas, polar bears, and whales.

DEFORESTATION AND PESTICIDES

The unchecked conversion of the world's forests, both tropical and temperate, for agricultural, commercial, mining, and residential development fragments these sensitive landscapes. Pollution of the air, soil, and water poses a serious threat to all organisms, including beetles. Pesticide drift—such as insecticides applied to fields in an effort to combat agricultural pests that are carried by wind or water to natural habitats—is especially harmful to shore-loving tiger beetles. The presence of the anti-parasitical ivermectin in cattle dung adversely affects the development rates and larval survival of beneficial dung beetles. The proliferation of companies needlessly applying pesticides to control mosquitoes kills all insects, including beetles. Wildfires, electric lights (including bug zappers), invasive species, water impoundments, off-road vehicles, and logging also threaten beetle populations.

← Conversion of land for residential, agricultural, and commercial development is the primary mechanism through which habitats are lost permanently.

CONSERVATION EFFORTS

The first step to conserving beetles and their habitats is to understand and appreciate their collective roles in the ecosystem upon which we depend for food, clean water, and breathable air. Functioning as herbivores, predators, and recyclers, beetles are not only essential to the sustainability of terrestrial ecosystems, but they are also incredibly useful as biological indicators and embody an enormous amount of scientific information. It just makes sense to invest in their conservation. Degrees of endangerment and endemicity, coupled with their ecological preferences and behavioral characteristics, especially in light of human activities, all provide conservation biologists with a unique set of challenges and opportunities to protect beetles.

Formally recognizing beetles as rare, threatened, or endangered transforms them into flagship species that raise public awareness of their plight and encourages financial support to conserve their habitats. Efforts to conserve a single species and its habitat results in the protection of all species in that habitat, a phenomenon known as the umbrella effect. Although beetles are among the most conspicuous and charismatic of all insects, the overall lack of knowledge of their biology, ecology, and distribution hampers efforts to identify and protect species in need of conservation. Thus, few species are recognized as needing conservation and afforded legal protection.

IDENTIFYING THREATENED SPECIES

The Red List of Threatened Species, or Red Data Book, was developed by the International Union for Conservation of Nature (IUCN) to provide comprehensive information on the extinction risk of organisms around the world. Listed species are included in one of eight categories: extinct, extinct in the wild, critically endangered, endangered, vulnerable, near threatened, least concern, and data deficient. It currently lists more than 1,700 beetle species, the vast majority of which are categorized as at least vulnerable or data deficient. Many countries have produced their own endangered species lists and enacted legislation to protect beetles and other wildlife by preserving and restoring their habitats, limiting development in areas known to be inhabited by listed species, and regulating or prohibiting their commercial exploitation.

LARGEST, LONGEST, SMALLEST

African Goliath beetles (*Goliathus*), Central and South American elephant beetles (*Megasoma*), and Atlas beetles (*Chalcosoma*) of Southeast Asia are all among the world's largest beetles. Tipping the scales at 3¹/₂ oz (99 g), the Guyanese *Megasoma actaeon* weighs more than a deck of playing cards.

In Central America, the bodies of male Hercules beetles (*Dynastes hercules*) measure only about 3 in (76 mm) in length but can reach up to a total of 7 in (178 mm) if you include their characteristic prothoracic horn. The longest-bodied beetle, aptly named Titan beetle (*Titanus giganteus*) from the Amazon rainforest, measures up to a whopping (6⁹/₁₆ in (175 mm) in length!

Measuring only ¹³/₁₀₀ in (0.325 mm), *Scydosella musawasensis* from Central and South America is the world's smallest beetle. In fact, this featherwing beetle is the world's smallest non-parasitic insect and could complete its entire life cycle within the Titan beetle's head with plenty of room to spare.

← Microscopic insects such as *Scydosella musawasensis* (Ptinidae) are of scientific interest because they are models for animal miniaturization studies.

← The larvae of Atlas beetles, *Chalcosoma atlas* (Scarabaeidae), eat rotten wood and take up to three years to complete their development.

→ The powerful mandibles of *Titanus giganteus* (Cerambycidae) can snap a pencil in two with ease. Despite its impressive size, little is known about this beetle's natural history other than that the adult males, not females, are attracted to lights.

Actual
size

2/5 in
(10 mm)

BLINDING SPEED

For its size, the fastest known running insect is a flightless Australian tiger beetle, *Rivacindela hudsoni*. Measuring ¾ in (20 mm) in length, it has been clocked running at top speeds of 8 ft (2.5 m), or 125 body lengths, per second for an average speed of 5.5 mph (9 km/h). This pales in comparison with the speed of the fastest human, Jamaican sprinter Usain Bolt, who set the world record for the 100-meter sprint at 9.58 seconds, which is 23.35 mph (37.58 km/h). However, if we scale speed for body length, at 6 ft 4¾ in (1.95 m) in height Bolt would have to run 207 miles (333.45 km) or 125 of his body lengths per second, reaching an average speed of 746 mph (1200.42 km/h)!

TIGER BEETLES

There are more than 2,600 species of tiger beetles living in diverse habitats around the world, save for the Arctic north of 65° latitude, Antarctica, Tasmania, and isolated oceanic islands such as Hawaii. They are often abundant along shorelines and on dry lakes and other saline habitats.

Tiger beetle larvae are unique among those of other beetles in that they dig narrow burrows and anchor themselves within using two pairs of upcurved abdominal hooks. They block the entrance to their burrows with their heavily armored and cryptically colored head and prothorax held flush with the soil surface. The predatory larvae ambush insects that venture too close to the burrow's entrance by lunging out backward and seizing them with their powerful mandibles, before dragging them down into their burrow.

← The bulging eyes, curved mandibles, and slender legs of adult tiger beetles (Cicindelidae) are adaptations for their predatory way of life.

Either diurnal or nocturnal, adult tiger beetles are leggy, fleet of foot, and are easily recognized by the bulging eyes and long, toothy, and sickle-shaped mandibles. Although some species are arboreal and live in trees, most are ground dwellers. Many species have specific habitat requirements and prefer to live on beaches, mudflats, dunes, sandy shorelines of streams and rivers, open forest floors, grasslands, boulder fields, or saline playas, to name a few.

Tiger beetles are visual hunters that either ambush their insect prey or pursue them at relatively high speeds in a curiously clipped fashion. The stop-and-go hunting style of terrestrial tiger beetles is a direct result of their speed hampering the ability of their eyes to gather enough photons. Photons, minute bits of electromagnetic energy that make up light, help tiger beetles to form a clear image of their prey. Although they possess excellent vision, tiger beetles accelerating at relatively high speeds toward their prey lose the ability to see and become temporarily blind. To compensate, they must stop several times for just a few milliseconds to gather enough photons reflecting off their prey in order to relocate it and resume the pursuit.

TOUCH AND GO

The staccato style of chasing prey displayed by tiger beetles is unusual in nature, and their particular way of dealing with temporary blindness is unique. Even though they may have to stop three or four times during the chase, the speed of tiger beetles still enables them to overtake their prey. While in pursuit, tiger beetles run with their antennae held rigidly out in front of them, not quite a couple millimeters off the ground. This behavior helps them to avoid running into any obstacles while they are chasing their prey. Tests performed in the laboratory have demonstrated that tiger beetles are indeed more dependent on their sense of touch, rather than sight, for avoiding physical obstacles in their environment.

SCRAPERS AND TAPPERS

To produce sound, some beetles will stridulate by rubbing a file-like structure against a roughened edge called a scraper. Beetles stridulate during courtship, in confrontations with other beetles, or in response to stress. Some species, such as hide and longhorn beetles, will stridulate by rubbing their legs or abdomens against their elytra to create chirping or squeaking alarms when attacked, possibly to startle predators. In burying and bess beetles, stridulation may help adults and their offspring to communicate with one another and stay within close proximity of each other, a hypothesis strengthened by the dependance of these larvae upon their parents for steady supplies of food, especially right after they hatch.

TAPPING BEETLES

Male and female deathwatch beetles (*Xestobium rufovillosum*) call to one another by tapping their heads against the walls of their wooden galleries. Females respond to calling males with taps of their own. A familiar sound on quiet summer evenings within aged oak timbers in old homes and other historic buildings in Europe, the persistent tapping sounds heard during silent bedside vigils have long been interpreted as signals of imminent death.

Toktokkies (including *Dichtha*, *Mariazofia*, and *Psammodes*) are iconic darkling beetles that inhabit forests, mountains, and deserts in southern Africa. Males and females engage in a form of sexual communication called substrate tapping. They call out to one another by rapidly drumming their abdomens against rocky substrates to produce a tapping sound described in Afrikaans as "*tok, tok*." An onomatopoeia, the moniker toktokkie is broadly applied to many other darkling beetles, whether they drum or not.

→ In 2022, Polish coleopterist Marcin Kamiński established *Mariazofia*, a genus of African toktokkie named in honor of his two daughters, Maria and Zofia.

ODONTOTAENIUS DISJUNCTUS

The horned passalus, *Odontotaenius disjunctus*, inhabits woodlands throughout eastern North America. Adults excavate tunnels in decayed logs where they live in subsocial family groups. The larvae rely on wood shredded and chewed by the adults as food, as well as the gut microbes associated with adult frass (excrement). Both adults and larvae produce sounds by stridulation. Adults call by rubbing small spines on the undersides of their flight wings across corresponding textured surfaces on the abdomen, while the larvae produce calls by rubbing their relatively small hind legs against the middle legs. Adults produce 14 different sounds that serve both social and defensive functions.

FEMMES FATALES

Fireflies and lightningbugs are neither flies nor true bugs, but soft-bodied beetles. Bioluminescent species emit a cold light as a form of sexual communication and defense. The male's light, which is species-specific in both pattern and duration, plays an important role in courtship by announcing his availability and location to females nearby. Males typically display their lights as they fly in search of females, while the usually stationary females respond with their own species-specific light signal.

BIOLUMINESCENT WARNING

Fireflies protect themselves from attack by spiders, birds, and other predators by emitting droplets of noxious hemolymph from their pronotum and elytra, a defensive behavior known as reflex bleeding. Their hemolymph is laced with distasteful steroids called lucibufagins. Although the amount of hemolymph emitted is substantial, fireflies appear to suffer no ill effects. Experienced vertebrate and invertebrate

predators soon associate bioluminescence with a bad taste and quickly learn to avoid eating fireflies. Thus, bioluminescence in fireflies is also a form of aposematism.

Female *Photuris* fireflies lack lucibufagins. To acquire this defensive chemical compound, they alter the pattern of their light signal to masquerade as a female *Photinus* firefly and lure in unsuspecting male *Photinus* as meals rather than mates. Once enriched with her victim's lucibufagins, the sated female *Photuris* switches her signal to attract a male of her own species, not as a meal, but as a mate. The lucibufagins acquired by these femmes fatales enhance their existing chemical defense system and will eventually be passed on to their eggs to protect them from predation by ants.

↙ *Photuris* females are known as "femmes fatales." By mimicking *Photinus* females, they lure *Photinus* males in order to eat them to acquire their defensive chemical, lucibufagins.

AGGRESSIVE MIMICRY

Female *Photuris* fireflies duping and eating male *Photinus* fireflies by imitating female *Photinus* is an example of aggressive mimicry. Another illustration of this phenomenon involves predatory *Elytroleptus* (Cerambycidae), which resemble their chemically protected prey, *Neolycus* (Lycidae), in the American Southwest and adjacent Mexico. *Elytroleptus apicalis* targets its lookalikes *Neolycus fernandezi* and *N. arizonensis*, while *E. ignitus* hunts its lycid doubles, *N. loripes* and *N. simulans*. Both species of *Elytroleptus* capture and consume *Neolycus* among their feeding and mating aggregations. Interestingly, neither predator co-opts their prey's feeding deterrent, lycidic acid, nor is their ability to tolerate this potent odorant understood.

BEETLE IMPERSONATORS

Other arthropods may resemble the appearance of beetles as a result of convergent evolution. Convergent evolution occurs when unrelated species occupy similar ecological niches and adapt to them in similar ways, morphologically and/or behaviorally, in response to similar selective pressures. Some insects and spiders share the aposematic patterns of toxic or otherwise chemically defended beetles.

MOTHS

Many species of tiger moths defend themselves with distasteful chemicals and are aposematically colored. In North America, the black-and-yellow lichen moth, *Lycomorpha pholus*, and the orange-patched smoky moth, *Pyromorpha dimidiata*, both have wings that strongly share the same aposematic color patterns of *Caenia*, *Calopteron*, and *Lycus* beetles in the family Lycidae. The wings of the painted lichen moth, *Hypoprepia fucosa*, resemble the colors and patterns of fireflies in the genera *Photinus* and *Photuris*.

COCKROACHES

Several species of tropical cockroaches resemble beetles in color and behavior. Species in the ectobiid cockroach genus *Prosoplecta* are more or less circular in outline and have short and thickened forewings that resemble elytra. Their beetle-like appearance is further reinforced by having relatively short antennae and legs, thus resembling beetles in several families, including jewel, lady, and leaf beetles. As its name suggests, the Brazilian firefly mimic, *Schultesia lampyridiformis*, a species often found in bird nests, is remarkably similar in appearance to fireflies. These and other beetle-like cockroaches are usually considered generalized mimics because their specific beetle models are mostly unknown.

OTHER FAKE BEETLES

Other apparent beetle impersonators have evolved armored structures that superficially resemble elytra. The mostly Old World family Celyphidae, commonly known as beetle flies, contains nearly 100 species that live primarily in Southeast Asia. Beetle flies are easily recognized by their enormously inflated scutellum, a portion of the thorax that covers most of the abdomen and shields the fly's single pair of wings when at rest. Whether this unusual morphological adaptation serves any additional functions is unknown.

Some true bugs resemble beetles because they, too, have abdomens that are partially covered by a scutellum-resembling elytra. True bugs are readily distinguished by having piercing-sucking, rather than chewing, mouthparts, and hemimetabolous development characterized by only three developmental stages: egg, nymph, and adult. The compact and beetle-like kudzu bug, *Megacopta cribraria*, an Old World plataspid, is now a widely established pest of soybeans and other legumes in the southeastern United States. Ebony bugs in the family Thyreocoridae also look like small beetles. They are broadly oval and shiny insects with a greatly enlarged scutellum that covers most of their abdomen and wings.

Earwigs are sometimes confused with rove beetles because of their slender bodies and short, thickened forewings call tegmina. However, these hemimetabolous insects are easily distinguished by the very conspicuous pair of pincher-like cerci located on the tip of their abdomen.

Insects aren't the only arthropods that resemble beetles. African and Asian orb-weavers in the genus *Paraplectana* are commonly called ladybird mimic spiders. The tough abdomen, lady beetle-shaped and -patterned, extends over the spider's cephalothorax like an Irish cap. Species of the South American orb-weaver genus *Encyosaccus* resemble leaf beetles. Birds that have eaten distasteful lady and leaf beetles will likely avoid eating them again, or any spiders that look like them.

↓ Flies in the family Celyphidae are known as beetle flies or beetle-backed flies. The scutellum, an enlarged plate on their thorax, functions like protective beetle elytra.

OF BEETLES AND BOTTLES

Among the most curious of Australian beetles is *Julodimorpha saundersii*. This large (1³/₈–2⁹/₁₆ in / 35–65 mm) and robust jewel beetle has golden-brown elytra that are distinctly sculpted with irregularly spaced punctures. They live in the mallee scrub habitats of southwestern Australia, where they are perhaps best known for the male's attraction to "stubby" beer bottles.

During the austral spring months of August and September, the fully winged males take flight and make a loud buzzing sound as they search for the flightless females. Driven by their preference for larger females, amorous males are sometimes attracted to discarded amber-colored beer bottles with pebble-like embossing. Scrambling over the bottles with their reproductive organs extended, the males apparently confuse the color and texture of these manufactured "evolutionary traps" with those same qualities of the female's elytra they find attractive. Other similarly colored and textured objects, especially orange peels, also attract the attentions of mate-seeking males.

↓ The only other species of *Julodimorpha*, *J. bakewellii*, occurs in eastern Australia and was depicted on a $2 Australian stamp in 2016.

↓ The single known larva of *J. saundersii* was discovered in a 7.2 ft (2.2 m) deep trench dug for a natural gas pipeline.

→ The attraction of *Julodimorpha saundersii* (Buprestidae) males to stubbies was first published in 1983 by Darryl Gwynne and David Rentz. In 2011, they were honored with an Ig Nobel Prize, a Nobel spoof for research that first makes people laugh, then think.

Stigmodera gratosa

$1
AUSTRALIA

GLOSSARY

aposematic
Distinctive, often contrasting color patterns that defensively warn predators of unpalatability or some other harmful characteristic.

bioluminescence
Light production involving oxidation of luciferin through the action of luciferase in select beetle families.

clypeus
A small *exoskeletal* plate at front of head that covers the mouthparts.

coleoptera
Holometabolous insect order commonly known as beetles and characterized by having chewing mouthparts and modified forewings called *elytra*.

diurnal
Active during the day.

dorsal
Back, upper side, or top.

ectoparasitoid
A parasitic larva that feeds externally on its host and ultimately kills it.

elytron (pl. elytra)
Modified leathery or shell-like forewing characteristic of beetles.

endoparasitoid
A parasitic larva that feeds internally within its host, ultimately killing it.

endosymbiotic microorganism
An organism that lives inside another.

exoskeleton
Protective outer covering of beetles that functions as both skeleton and skin; serves internally as a foundation for muscles and organ systems, while providing a platform for sensory structures, appendages, and wings externally.

fungivore
Feeds on fungus; mycophage.

hemolymph
Bodily fluid of insects that functions as both blood and lymph in vertebrates.

holometaboly
Development with four distinct stages (egg, larva, pupa, adult); also called complete metamorphosis.

hydrofuge
Dense, water-repellent setae found on the *exoskeletal* surface of aquatic beetles.

hypermetamorphosis
Holometabolous development usually found in *parasitic* beetles (Bothrideridae, Meloidae, Rhipiceridae, Ripiphoridae) where larval stages differ in form.

larviform
Adult female beetles lacking wings and resembling larvae; distinguished by having compound eyes and fully developed reproductive organs.

mycangia (sing. mycangium)
Exoskeletal receptacles in bark beetles that carry *symbiotic* fungi.

ovipositor
Abdominal structure in female insects that facilitates oviposition, or egg laying.

parasite
An organism dependent on another organism or host for its existence; usually does not kill host.

parasitoid
A parasite that typically kills its host.

parthenogenetic
An organism capable of parthenogenesis, or reproduction without fertilization.

plastron respiration
Respiration method used by some aquatic beetles in which a thin layer of air is trapped by a velvety mesh of dense *pubescence* surrounding the body (plastron) and used to obtain dissolved oxygen from surrounding water and expel carbon dioxide.

pubescence
Soft, fine, short, loosely set, and erect *setae*.

punctures
Small pits on surface of *exoskeleton*.

pygidium
Last dorsal abdominal sclerite (tergite) in beetles.

reflex bleeding
A defensive release of *hemolymph* through intersegmental membranes between leg joints and body segments.

riparian
A narrow band of woodland flanking streams and rivers.

saproxylic
An organism that inhabits dead or decaying wood.

scale
A flattened *seta*, ranging in outline from nearly round, to oval (egg-shaped), obovate (pear-shaped), lanceolate (spear-shaped), or linear (long and slender).

seta (pl. setae)
Hairlike structure of insects.

spermatheca
Female insect organ that stores and nourishes sperm until fertilization and oviposition.

spiracle
External opening of tracheal system.

stridulate
Rubbing one body part against another, usually by moving a row of spines or small bumps across a ridge or series of ridges to produce sound.

subelytral cavity
Space beneath the *elytra* used by aquatic beetles in which to store air in order to breathe under water; also aids in thermoregulation of terrestrial species.

symbiotic
In reference to different species living in association with another, but does not imply the nature of the relationship.

teneral
Freshly enclosed pale and soft-bodied adult insect.

thanatosis
Playing dead as a defensive tactic so that predators lose interest.

triungulin
Active first larval stage of a *parasitic* beetle that develops by *hypermetamorphosis*.

ventral
Located below or relating to the underside.

FURTHER READING

Barclay, M.V.L. and P. Bouchard. 2023. *Beetles of the World. A Natural History*. Princeton, NJ: Princeton University Press.

Bouchard, P. (ed.). 2014. *The Book of Beetles: A Life-size Guide to Six Hundred of Nature's Gems*. Chicago: University of Chicago Press.

Evans, A.V. 2014. *Beetles of Eastern North America*. Princeton, NJ: Princeton University Press.

Evans, A.V. 2021. *Beetles of Western North America*. Princeton, NJ: Princeton University Press.

Evans, A.V. 2023. *The Lives of Beetles: A Natural History of Coleoptera*. Princeton, NJ: Princeton University Press.

Marshall, S. 2018. *Beetles: The Natural History and Diversity of Coleoptera*. Richmond Hill, ON, Canada: Firefly Books.

INDEX

ACKNOWLEDGMENTS

The Little Book of Beetles was very much a collaborative effort that began with an invitation from UniPress' Nigel Browning to write this book. As with my previous project with UniPress, *The Lives of Beetles*, it was truly a pleasure to collaborate once again with Ruth Patrick. Her patience and good humor during the production of this book helped immensely to keep everything on track. Tugce Okay's lavish and detailed illustrations regularly added shape, texture, and color to my prose. Iain Durneen's charming incidental monotone illustrations sprinkled throughout the book provided additional context. I am especially grateful to John Abbott and Joyce Gross for permission to use their fine images in this book. Lindsey Johns artfully arranged all of these elements into a very unique and attractive little tome. I am especially thankful to our literary partner, Princeton University Press, where Robert Kirk and his colleagues Megan Mendonca and Ruthie Rosenstock continue to be wonderfully supportive of my work.

The book's essays were drawn largely from primary sources, particularly the Biodiversity Heritage Library (biodiversitylibrary.org) and Journal Storage (jstor. org). Special thanks to the Smithsonian Libraries and Archives in Washington, DC and Boatwright Memorial Library, University of Richmond in Virginia, for providing access to other electronic journals. Additional publications consulted during the research phase to this book were generously made available by their authors at ResearchGate (researchgate.net).

Robert Anderson of the Canadian Museum of Nature kindly read portions of an early draft of the manuscript, then reviewed the entire book. An equally gifted coleopterist and science communicator, his cogent remarks and suggestions greatly improved the accuracy and readability of this book.

My wife Paula Evans enthusiastically read an early draft of the book and offered numerous suggestions that helped to improve its overall clarity and readability. Her love and support over the past quarter century remain a constant source of joy and comfort to me.

As with all my literary efforts, I share the success of this work with all the aforementioned individuals, but any and all of its shortcomings are solely my own.

ABOUT THE AUTHOR

Arthur V. Evans is an entomologist, educator, photographer, radio broadcaster, and Emmy Award-winning video producer. His many books include *Beetles of Western North America* and *The Lives of Beetles: A Natural History of Coleoptera* (both Princeton). He lives in Richmond, Virginia.

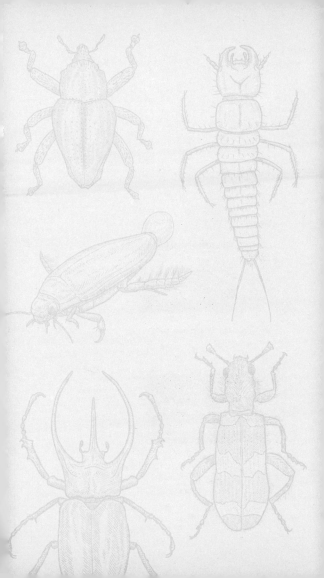

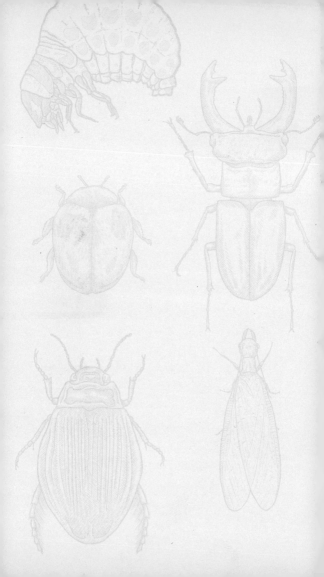